RESOURCES FOR THE HISTORY OF PHYSICS

RESOURCES FOR THE HISTORY OF PHYSICS

I. Guide to Books and Audiovisual Materials

II. Guide to Original Works of Historical Importance and Their Translations into Other Languages

Edited by: Stephen G. Brush
University of Maryland

The University Press of New England

Hanover, New Hampshire

1972

PHYSICS

Published for the International Working Seminar on the Role of the History of Physics in Physics Education, A Conference held at the Massachusetts Institute of Technology from 13 - 17 July 1970 and financed by the United Nations Educational, Scientific, and Cultural Organization, the Alfred P. Sloan Foundation, and the National Science Foundation.

Library of Congress Catalogue Card Number: 70-186306.
International Standard Book Number: 0-87451-066-X.
Printed in the United States of America.

UNESCO Subvention - 1970 - DC/2.1/414/28

I

GUIDE TO BOOKS AND AUDIOVISUAL MATERIALS

PREFACE

The historical approach to science probably is as old as science itself; however, it has never been the only approach nor should it be. The seminar that led to the preparation of this guide was a response to several contemporary trends which have combined to focus attention on the role of the history of physics in physics education. First, many physics teachers have become convinced that some appreciation of the historical development of physical science is essential to understanding its present nature and function; historical perspective can be valuable both in seeking new knowledge and in applying existing knowledge. Second, in an era when science is becoming ever more highly specialized and the specialists less and less familiar with work outside of their narrow disciplines, perhaps the fact of common origins made vivid in the history of science can contribute something to holding the whole scientific enterprise together; the history of science forms a common interest of people in many diverse specialties. Third, the historical approach has been found especially useful in teaching physics to the general student who is more interested in the personalities, philosophical aspects, and social role of science than in technical details. Fourth, the recent flowering of history of science as a professional discipline has provided the teacher with an ample supply of literature and, in many cases, with faculty colleagues trained in the field. It therefore seemed appropriate to bring together in a small seminar a group of physics teachers and historians of science for the purpose of assessing the present situation and exploring new directions. The International Commission on Physics Education, therefore, appointed an Organizing Committee, with Professor Allen L. King of Dartmouth College as chairman, to carry out the project. The proceedings of the seminar may be found in _History in the Teaching of Physics_, edited by Brush and King and published by the University Press of New England. One product of the assessment and exploration undertaken by the seminar is this book, _Resources for the History of Physics_.

Many teachers who wish to use historical materials often lack the time, background, or inclination for historical scholarship and, therefore, limit themselves to the superficial (and often inaccurate) information that can be gleaned from textbooks and popular works on the history of science. But there is available now a wealth of materials which can bridge the gap: monographs on special topics or periods in the history of science, journal articles and case studies, biographies, and films. We hope that the following guide to these materials will be a useful tool despite the fact that it really cannot be either complete or up to date. (We welcome corrections and suggestions of other items for inclusion in future editions.) The present list includes information supplied by several participants in the

International Working Seminar on the Role of the History of Physics in Physics Education. In addition, many items were suggested by Dr. M.G. Ebison of the College of St. Mark and St. John, London. Comments and evaluations are by the editor unless otherwise attributed. The final editing and retyping of the text for reproduction have been accomplished under the supervision of Professor King; his assistance has been of immense value at all stages in the preparation of the book. We are especially grateful to Mrs. Donna Musgrove for her patience and skill in producing the typed copy ready for the photo-offset printing process.

The numbers at the right of each item indicate the age group for which it primarily is intended; or else there is an indication that it is more suitable for teachers (T) or libraries (L). For some items we have omitted such evaluations for lack of information on the matter. Information about the books and films generally is that given in the 1970 edition of Books in Print, in the National Union Catalogue, or in publishers' catalogues. Prices have been omitted because they are subject to change; the publishers should be queried on the matter. We have indicated whether a book is published as a paperback by (P); but often such a book has been published in hardcover too. Although some excellent books are out of print (O/P), they are included in the list because they can be found in libraries or may be obtained secondhand.

On behalf of the International Commission on Physics Education, the Organizing Committee, and the participants in the seminar we gratefully acknowledge the financial support of the Alfred P. Sloan Foundation, the National Science Foundation, UNESCO, and the International Union of Pure and Applied Physics.

S.G. Brush

January 1972

Part I

TABLE OF CONTENTS

Part I

Page

1. GENERAL BOOKS ON THE HISTORY OF SCIENCE: (a) ENGLISH

Bernal, J.D. Science in History. 3d ed.; London: Watts, 1965; New York: Hawthorn Books, 1965. (L)

An attempt to describe and interpret the relations between the development of science and other aspects of human society. The book mostly concerns scientific development from earliest times to the present. The final part deals with the interrelations between science and politics. (B. Gee)

Collingwood, R.G. The Idea of Nature. London: Oxford University Press, 1945. (18-21)

An examination of the idea of Nature in three great periods of European thought: Greek, Renaissance and Modern. (M.G. Ebison)

Crowther, J.G. A Short History of Science. London: Methuen Educational, 1969. (16-18)

This book traces the development of science as one of the major responses of man to his experience of living.

Dampier, W.C. A History of Science and Its Relations with Philosophy and Religion. 5th ed.; Cambridge: University Press, 1966. (P) (T,L)

This edition has a postscript by I.B. Cohen containing a bibliography of recent works. There is also a Shorter History of Science by the same author. (Meridian World)

German translation by F. Ortner. Wien: Humbolt Verlag, 1952. Ist nicht international genug und z. Teil oberflächlich. Als einführender Überblick brauchbar. (H. Kangro)

Gillispie, C.C. The Edge of Objectivity. Princeton, N.J.: Princeton University Press, 1960. (P) (20-24,T)

From the natural philosophy of the 17th century through the biology and field physics of the 19th century.

Goatley, James L. The Search for Explanation, Studies in Natural Science. Vol. 1. Ann Arbor: Michigan State University Press, 1967. (O/P)

A textbook history of the physical sciences.

Hull, L.W.H. History and Philosophy of Science. London: Longmans, 1959. O/P in U.S.A. (18-21, T)

This stylishly written book traces the growth of scientific ideas from Greek to modern times against a general background of history. The connections between science and other models of thought is particularly emphasized throughout the book. (M.G. Ebison)

Kuhn, Thomas S. The Structure of Scientific Revolutions. 2d ed.; Chicago: University Press, 1970. (19-22, T)

A very influential book in the contemporary development of the philosophy of science. The author stresses the importance of "revolutions" in science. (M.G. Ebison)

Mason, S.F. Main Currents of Scientific Thought: A History of the Sciences. London: Routledge, 1956. Rev. ed. A History of the Sciences. Collier Books, 1962. (18-22)

An excellent general history of the sciences. After a brief treatment of ancient and medieval science, most of the book is concerned with the development of modern science from the sixteenth century onwards. Bibliography. (B. Gee)

In a recent survey conducted by the United States Armed Forces Institute, this was found to be the most frequently adopted textbook for American college courses in the history of science. (S. Brush)

McKenzie, A.E.E. The Major Achievements of Science. 2 vols. Cambridge: University Press, 1960. (15-18)

Volume 1 deals with the historical development of some of the major concepts and topics in science, and five chapters are interspersed which deal with the main trends from ancient to modern times. Volume 2 contains 91 brief extracts from scientific literature to accompany volume 1. Specially written for secondary school education. (B. Gee)

Pledge, H.T. Science Since 1500. 2d ed.; London: H.M.S.O., 1966. (HP)

A valuable reference work containing a wealth of information on the history of mathematics, physics, chemistry and biology. (M.G. Ebison)

Ronan, C.A. The Ages of Science. London: Harrap, 1966. (16-18)

Beginning in the earliest times, the author shows how theories and ideas of every age seemed in their day so logical and so well in accord with the facts then known. (M. B. Ebison)

Shepherd, Walter. Outline History of Science. London, Melbourne: Ward & Lock Educnl., 1965. New York: Philosophical Library, 1968. O/P in U.S.A. (T)

This book provides a rapid summary of notable events in science from Babylonian times to 1965, arranged in chronological order. There are also lists of "Notable Inventions and Discoveries," "Laws and Principles," and "Outstanding Scientists." It is a useful book for every teacher to have despite one or two inaccuracies. (B. Gee)

Singer, C. A Short History of Scientific Ideas to 1900. London: Oxford University Press, 1963. (P) (18-21, T)

This book presents the development of the concept of a material world in which all parts are rationally interrelated.

Steele, D., ed. History of Scientific Ideas, A Teacher's Guide. London: Hutchinson Educational, Ltd., 1970. (P) (T)

To help teachers to run courses, at schools and in technical institutions, in the History of Science. Contributors include Tony Joyce, David Newbold, and David Hughes Evans. (Blackwell's of Oxford catalog)

Taylor, F. Sherwood. Science: Past and Present. London: Heinemann, 1945; Mercury Books, 1962. (16-18)

The author attempts to depict science as a growing organism with powers and limitations. The author supplies collections of extracts from contemporary scientific works for each chapter in the book. (M.G. Ebison)

Whewell, William. History of the Inductive Sciences. London: Frank Cass, 1967. Reprint of the 1857 edition.

Wiener, P.P. & Noland, A. The Roots of Scientific Thought. New York: Basic Books 1956. (L)

This book presents articles on the development of the world's dominant scientific ideas by 30 scholars from the United States, Canada and Europe. It is divided into four parts: the classical heritage, from rationalism to experimentalism, the scientific revolution, from the World-machine to cosmic evolution. (M.G. Ebison)

Wightman, W.P.D. The Growth of Scientific Ideas. London: Oliver and Boyd, 1966. New Haven: Yale University Press, 1951, 1964. (T, L)

By consideration of a few topics the author shows some developments in scientific thinking. The Copernican system, mechanics and gravitation, theories of light and colour, phlogiston, caloric, electricity, energy, etc. There is also some treatment of the biological sciences. (B. Gee)

1. GENERAL BOOKS ON HISTORY OF SCIENCE: (b) FRENCH

Daumas, M., ed. Histoire Générale des Techniques. Paris: Presses Universitaires de France, 1962.

Les origines de la civilisation technique par C. de La Calle, et al.

Hooijkaas, R. L'Histoire des Sciences, ses Problemes, sa Méthode, son But. Coimbra: 1963.

Russo, F. Histoire de la Pensée scientifique. Paris: La Colombe, 1951.

Taton, R., ed. Histoire générale des Sciences. Paris: Presses Universitaires de France. 4 vols. 1957-62. (L) (Also available in English translation.)

I. Ancient and Medieval Science. II. Modern Science (15th to 18th centuries). III. 19th century. IV. 20th century.

This book is at present the only one of its size and coverage available in the French language. Most of its contributors write for the cultured general reader rather than for any specialist. Sources and statement of origin are usually taken secondhand. (G.A. Boutry)

1. GENERAL BOOKS ON HISTORY OF SCIENCE: (c) GERMAN

Bernal, J.D. Die Wissenschaft in der Geschichte. Trans. from English. Berlin: Progress-Verlag, 1967. (See Section la.)

Dannemann, Friedrich. Grundriss einer Geschichte der Naturwissenschaften, zugleich eine Einführung in das Studium der grundlegenden naturwissenschaftlichen Literatur. 1. Bd., Leipzig: W. Engelmann, 1896 (2. Aufl. 1902); 2. Bd., 1898. 3. Aufl. unter dem Titel: Aus der Werkstatt grosser Forscher. Allgemeinverständliche erläuterte Abschnitte aus den Werken hervorragender Naturforscher aller Völker und Zeiten. Leipzig: W. Engelmann, 1908. (16-24)

> Die Zeit von Aristoteles bis H. Hertz wird an Hand von vielen, gut ausgewählten Originalquellen abgehandelt. Noch immer also getreu dokumentierte Geschichte brauchbar. (H. Kangro)

Dampier, W.C. A History of Science (see Section la).

Mason, S.F. Geschichte der Naturwissenschaft in der Entwicklung ihrer Denkweisen. Trans. from English by K. Meyer-Abich. Stuttgart, 1961. (See Section la.) (18-22)

> Eine mit seltener didaktischer Treffsicherheit und vorzüglichem historischem Gefühl für Geisteszusammenhänge dargestellte Geschichte; Astronomie und Biologie werden vor Physik und Chemie bevorzugt. Nur Dokumentation der Sekundärquellen. (H. Kangro)

Schimank, Hans. Epochen der Naturforschung, Leonardo, Kepler, Faraday. Berlin: Wegweiser-Verlag GMBH, 1930; 2. Aufl., Munchen: Heinz Moos-Verlag, 1964. (18-24)

> Geschickte an den Unterschieden charakterisierte Geschichte, mit feinem Einfühlungsvermögen nach gründlichem Quellenstudium dargestellt. (H. Kangro)

1. GENERAL BOOKS ON HISTORY OF SCIENCE: (d) JAPANESE

Forbes, R.J., and Dijksterhuis, E.J. _Kagaku to Gizyutu no Rekisi_. Trans. of _A History of Science and Technology_, 1963. 2 vols. Tokyo: Mixuzu Shobo, 1963-64.

Mason, S.F. _Kagaku no Rekisi_. Trans. of _A History of Science_, 1953. 2 vols. Tokyo: Iwanami Syoten, 1955-56.

Singer, Charles. _Kagaku-siso no Ayumi_. Trans. of _A Short History of Scientific Ideas to 1900_, 1959. Tokyo: Iwanami Syoten, 1968.

1. GENERAL BOOKS ON HISTORY OF SCIENCE: (e) OTHER LANGUAGES

Dijksterhuis, E.J., and Forbes, R.J. _Overwinning door gehoorzaamheid_. Antwerp: De Haan (Phoenix Pockets, nr 54 and 55), 1961. (In Dutch)

2. GENERAL BOOKS ON HISTORY OF PHYSICS: (a) ENGLISH

Agassi, Joseph, and Agassi, Aaron. The Continuing Revolution, A History of Physics from the Greeks to Einstein. New York: McGraw-Hill, 1968. 225 pp. (11-16)

Presented in the form of a father-son dialogue.

Amaldi, G. The Nature of Matter: Physical Theory from Thales to Fermi and After. London: Allen and Unwin, 1966. (18-21)

A very short introduction to the speculations of the Greek philosophers is followed by a detailed but readable description of the long path to the present understanding of atomic physics. (M.G. Ebison)

Cajori, Florian. A History of Physics. New York: Dover Publications, 1962. (P) London: Constable. (T, L)

The subject of physics is considered in its elementary branches and the scope of the work covers all aspects of physics up to 1925. It has an interesting section on the development of physical laboratories. The book was originally published in 1929 but the style is still unbeatable for its clarity in which the author has presented his work. For 1975 someone ought to add a supplement to this on the history of the past 50 years! (B. Gee)

Brauchbarer, gut ausgewählter Überblick, teilweise aus veralteten Quellen dargestellt. (H. Kangro)

Chalmers, T.W. Historic Researches. London: Morgan Bros., 1949. New York: Fernhill House, 1968. (T, L)

Fraser, Charles G. Half-Hours with Great Scientists, the Story of Physics. New York: Reinhold Pub. Co.; and Toronto: University Press, 1948. O/P in U.S.A.

Gamow, George. Biography of Physics. New York: Harper, 1961; Harper Torchbook. (P) (18-22)

A lively and entertaining book for the non-specialist student. (B. Gee)

German trans. by D. and D. Müller. Düsseldorf & Wien: Econ-Verlag, 1965.

Z. Teil oberflächlich und unhistorisch. (H. Kangro)

Laue, Max von. History of Physics. Trans. from German. New York: Academic Press, 1950. O/P in U.S.A. (19-22)

Lipson, H.S. The Great Experiments in Physics. Edinburgh: Oliver and Boyd, 1968. (P)

This well-written book is both lucid and authoritative. It is one of a very interesting series of books called Contemporary Science Paperbacks. (M.G. Ebison)

Schneer, Cecil J. The Search for Order. New York: Harper, 1960. Reprinted as The Evolution of Physical Science. New York: Grove Press, 1964. (P) 398 pp. (19-22)

A fascinating survey of the development of scientific ideas and methods from Ancient Greece to the present day. (M.G. Ebison)

Shamos, M.H. Great Experiments in Physics. New York: Holt, Rinehart & Winston, 1965. (P) (T, L)

Originally designed for a physics course at Washington Square College, this contains many original papers and extracts from works along with a commentary in a flying column. (B. Gee)

Snyder, Ernest E. History of the Physical Sciences. Columbus, Ohio: C.E. Merrill, 1969. (P) (18-20)

A textbook with review questions; covers a wide range of topics rather superficially.

Whittaker, Edmund. From Euclid to Eddington, A Study of Conceptions of the External World. Cambridge: University Press, 1947; New York: Dover Publications, 1958. (P) O/P in U.S.A. German trans., Von Euklid zu Eddington, Zur Entwicklung unseres modernen physikalischen Weltbildes. Vienna: 1952. (T)

2. GENERAL BOOKS ON HISTORY OF PHYSICS: (b) FRENCH

Brunold, Ch. _Histoire abrégée des Théories physiques concernant la Matière et l'Énergie_. Paris: Masson, 1952.

Marie, M. _Histoire des Sciences mathématiques et physiques_. Paris: 1887-1888. (L)

Obsolete in many places but still a useful book to have on the shelf. (G. Boutry)

Massain, R. _Physique et Physiciens_. Paris: J. de Gigord, 1950. ca. 350 pp. (17-19)

Mainly biographical; well adapted to finishing high school pupils and to sophomore students; unpretentious and easy to read. (G. Boutry)

Volkribger, H. _Les Étapes de la Physique_. Paris: 1929.

A shorter history of physics, mainly in the 19th and 20th centuries. (G. Boutry)

2. GENERAL BOOKS ON HISTORY OF PHYSICS: (c) GERMAN

Gerland, E., und v. Steinwehr, H. Geschichte der Physik, erste Abteilung: Von den altesten Zeiten bis zum Ausgange des achtzehnten Jahrhunderts (Geschichte der Wissenschaften in Deutschland, neuere Zeit: 24 Band). München & Berlin: R. Oldenburg, 1913. (18-25, T)

Mit gutem Gefühl für historische Zusammenhänge dargestellte Geschichte. Sorgfältig auf den bis 1913 erreichbaren Quellen aufgebaut. Enthält den nötigen historischen Hintergrund. (H. Kangro)

Gerland, E., und Traumüller, F. Geschichte der physikalischen Experimentierkunst. Hildesheim: G. Olms, 1965 (Neudruck der Ausgabe Leipzig, 1899).

Einmaliges Thema: Das Buch ist noch immer gut brauchbar. (H. Kangro)

Heller, A. Geschichte der Physik von Aristoteles bis auf die neuste Zeit, 2 Bände. Stuttgart: 1882-1884; Wiesbaden: M. Sändig, 1965. (18-24, T)

Wenn auch in einigen historischen Tatsachen veraltet, doch eine—gegen Hoppes Werk—mit vorzüglichem Sinn für historische Zusammenhänge dargestellte Geschichte. (H. Kangro)

Hoppe, Edmund. Geschichte der Physik. Braunschweig: F. Vieweg u. Sohn, 1926, 1965; New York: Johnson Reprint Corp., 1965. (T, L)

Ausserordentlich reiche, fast immer zuverlässige Materialsammlung, bewusst ohne Betrachtung irgendeines historischen Hintergrundes dargestellt. Bestimmte Autoren (z.B. Euler) werden einseitig bevorzugt. Zeit: Altertum bis ca. 1894-1895. Die Kapitel sind nach Sachgebieten aufgegliedert. Auch Sachverzeichnis (jedoch nur moderner Leitbegriffe) vorhanden. Unentbehrlich als Quellenwerk für die Geschichte der Physik. (H. Kangro)

Hunger, Edward. Von Demokrit bis Heisenberg. 3rd ed.; Braunschweig: F. Vieweg u. Sohn, 1963.

Laue, Max von. Geschichte der Physik. Frankfurt, 1947; 4th ed.; Bonn: Athenäum-Verlag, 1958. (T)

Kurzer geschichtlicher Rückblick, anregend, doch ohne historisches Gefühl. (H. Kangro)

Poggendorff, J.C. Geschichte der Physik. Leipzig, 1879; Leipzig: Central Antiquariat D.D.R., 1964.

Ramsauer, C. Grundversuche der Physik in historischer Darstellung, 1. Bd.: Von den Fallgesetzen bis zu den elektrischen Wellen. Berlin, Göttingen, Heidelberg: Springer, 1953. (17-20)

Weiter Bände sind nie erschienen. Es handelt sich um eine moderne Anleitung für die Wiederausführung von historischen Versuchen, historisch kompetent. (H. Kangro)

Rosenberger, F. Die Geschichte der Physik in Grundzügen mit synchronistichen Tabellen der Mathematik, der Chemie und beschreibenden Naturwissenschaften sowie der allgemeinen Geschichte. Bd. I-III in 2 Bänden, Hildesheim: G. Olms, 1965 (Neudruck der Ausgabe Braunschweig 1882-1890). (20-24, T)

Behandelt den Zeitraum vom Altertum bis 1880. Trotz des Fehlens der Physikgeschichte ab 1800 und trotz einiger weniger veralteter historischer Ansichten noch immer wegen der vorzüglichen Darstellung der Zusammenhänge recht brauchbar. (H. Kangro)

Hermann, Armin, ed. Lexikon der Schulphysik. Band 6 (A-K), Band 7 (L-Z). Köln: Aulis Verlag Deubner, 1971.

2. GENERAL BOOKS ON HISTORY OF PHYSICS: (d) JAPANESE

Cajori, Florian. Buturigaku no Rekisi. Translated from A History of Physics, 1929. 3 vols.; Tokyo: Tokyo Tosyo, 1964-1965.

Hirosige, T. Buturigaku-sî [A History of Physics]. 2 vols.; Tokyo: Baihukan, 1968.

Describes the development of modern physics since its genesis in the 17th century to the birth of quantum mechanics.

2. GENERAL BOOKS ON HISTORY OF PHYSICS: (e) OTHER LANGUAGES

Pannekoek, A. De Groei van ons Wereldbeeld, een Geschiedenis van de Sterrekunde. Amsterdam: Wereld-Bibliotheek, 1951. 440 pp. (In Dutch)

3. PHYSICAL SCIENCE IN RESTRICTED HISTORICAL PERIODS: ANCIENT AND MEDIEVAL

Carteron, H. La Notion de Force dans le Système d'Aristote. Paris: J. Vrin, 1923. (L)

Clagett, Marshall. Greek Science in Antiquity. New York: Collier Books, 1963. (19-22)

_______. Giovanni Marliani and Late Medieval Physics. New York: AMS Press; reprint of the 1941 ed. (L)

Crombie, A.C. Augustine to Galileo. London: Heinemann Mercury, 2nd ed., Medieval and Early Modern Science. Cambridge, Mass.: Harvard University Press, 1959; Doubleday Anchor Paperback. Japanese trans.: Tyusei kara Kindai eno Kagakusi. Tokyo: Corona, 1962, 1968. German trans. by H. Hofmann and H. Puls. Von Augustin bis Galilei, Die Emanzipation der Naturwissenschaften. Köln & Berlin: Kiepenheuer u. Witsch, 1964. (T, L)

> Volume 1 covers from the 5th to the 13th century; Volume 2 from the 13th to 17th centuries. A definitive work on medieval science, which includes excellent accounts of the mechanics, astronomy, physics, chemistry, geology, biology, and technology. Extensive bibliographies. (B. Gee)

Dijksterhuis, E.J. De Mechanisering van het Wereldbeeld (see Section 4).

Duhem, Pierre. Études sur Léonard da Vinci: ceux qu'il a lus; ceux qui l'ont eu. 2 vols.; Paris: Hermann, 1906-1909. (L)

Maier, A. Studien zur Naturphilosophie der Spätscholastik, I - V. Roma: Edizioni die Storia e Letteratura, 1949-1958. (T)

- I. Die Vorläufer Galileis im 14. Jahrhundert (1949);
- II. Zwei Grundprobleme der scholastischen Naturphilosophie: Das Problem der intensiven Grösse; die Impetustheorie (1951, 2. Auflage);
- III. An der Grenze von Scholastik und Naturwissenschaft: Die Struktur der materiellen Substanz; das Problem der Gravitation; die Mathematik der Formlatituden (1952, 2. Auflage);
- IV. Metaphysische Hintergründe der spätscholastischen Naturphilosophie (1955);

V. Zwischen Philosophie und Mechanik (1958).

Wegen der subtilen Behandlungsmethode auf Grund vorzüglicher Quellenkenntnis sind diese Schriften unentbehrlich, wenngleich sie ein wenig vom Standpunkt des engen Begriffes moderens Philosophie abgehandelt sind, was nur einen Aspekt des weiteren mittelalterlichen Begriffes der Wissenslehre 'philosophia' darstellt. (H. Kangro)

Mansion, A. Introduction á la Physique aristotelicienne. 2nd ed.; Louvain-Paris: 1945. (L)

Neugebauer, O. The Exact Sciences in Antiquity. 2nd ed.; Princeton: University Press, 1957; New York: Dover. (P) (T, L)

Sambursky, S. The Physical World of the Greeks. Trans. from Hebrew by Merton Dagut. 2nd ed.; London: Routledge and Paul, 1960; New York: Collier-Macmillan, 1962. (P) (T, L)

________. Physics of the Stoics. Edinburgh: Constable & New York: Humanities Press, 1959.

________. The Physical World of Late Antiquity. London: Routledge and Paul, 1962; New York: Basic Books, 1962. O/P in U.S.A. German translation, Das Physikalische Weltbild der Antike. Zürich and Stuttgart: 1965. (T, L)

Solmsen, Friedrich. Aristotle's System of the Physical World. A Comparison with His Predecessors. Ithaca: Cornell University Press, 1960. 468pp. O/P in U.S.A. (T, L)

van der Waerden, B.L. Science Awakening. Translated from Dutch by A. Dresden. Groningen, Holland: Noordhoff, 1954; New York: Wiley, Science editions, 1963; Oxford and New York: Oxford University Press. (T)

Wieland, W. Die Aristotelische Physik. Göttingen: 1962. (L)

4. PHYSICAL SCIENCES IN THE 16th, 17th & 18th CENTURIES

Boas [Hall], Marie. The Scientific Renaissance 1450-1630. London: Collins; and New York: Harper & Row, 1962. (P) (T, L)

A brilliant account of the changes in scientific attitudes from the mid-15th century when science was dominated by the recovered Greek writings, through the revolutionary theories of the 16th century when the new theories were beginning to be accepted. (B. Gee)

Brunet, Pierre. Les physiciens Hollandais et la Methode experimentale en France au XVIII siecle. Paris: Librairie Scientifique Albert Blanchard, 1926. (T, L)

_______. L'Introduction des Theories de Newton en France au XVIII siecle, I, Avant 1738. Paris: Librarie Scientifique Albert Blanchard, 1931. (T, L)

Burtt, E.A. The Metaphysical Foundations of Modern Physical Science. 2nd ed.; Garden City: Humanities Press, 1952; New York: Doubleday & Co., 1954. (P) (20-24)

An excellent survey and analysis of the philosophical aspects of the discoveries of Copernicus, Kepler, Galileo, Descartes, Boyle, and Newton.

Butterfield, Herbert. Origins of Modern Science. London: G. Bell, 1957; Collier Books, 1962; Free Press Paperback, 1965. (19-22)

Shows the relation of science to other developments in the 16th-18th centuries from the perspective of a general historian. In a recent survey conducted by the United States Armed Forces Institute, this was found to be the second most frequently adopted textbook for American college courses in the history of science, next to Mason's History of the Sciences.

Butts, R.E. and J.W. Davis, eds. The Methodological Heritage of Newton. Toronto: University Press, 1969. (T, L)

An excellent collection of essays by distinguished authors which is highly recommended. The essays assess Newton's contributions to the thought of others. (M.G. Ebison)

Cohen, I.B. *The Birth of a New Physics*. Garden City, N.Y.: Doubleday & Co., Anchor Books (available to secondary school students and teachers through Wesleyan University Press, Columbus, Ohio), 1960. 200pp. (P) London: Heinemann, 1960. (17-20)

Readable and authoritative account, emphasizing the astronomical background of Newtonian physics.

________. *Franklin and Newton, An Inquiry into Speculative Newtonian Science and Franklin's Work in Electricity As An Example Thereof*. Philadelphia: American Philosophical Society, 1956; Harvard University Press. 657pp. (L)

Daumas, M. *Les Instruments Scientifiques aux XVIIème et XVIIIème Siècles*. Paris, 1953. 417pp.

Dijksterhuis, E.J. *De Mechanisering van het wereldbeeld*. Amsterdam: J.M. Meulenhoff, 1950. German trans.: *Die Mechanisierung des Weltbildes*. Berlin, 1956. English trans.: *The Mechanization of the World Picture*. New York and London: Oxford University Press, 1961. (P) (21-25, T)

This is a comprehensive analytical account of the transition from the ancient (Aristotelian, Platonic) world views to the mechanical philosophy established in the 17th century. It is an excellent book but demands considerable intellectual effort from the reader. (S. Brush)

Eine ausgezeichnete Darstellung der Geschichte naturwissenschaftlicher Ansichten. Behandelt werden alle mechanistisch darstellbare Themen, von den frühen Griechen bis einschliesslich Newton. Die Tatsachen sind mit der notwendigen historischen Sorgfalt und Vorsicht beschrieben und auf gründliche Quellenkenntnis aufgebaut. Der ideengeschichtliche Hintergrund ist vortrefflich herausgearbeitet. (H. Kangro)

Hall, A. Rupert. *From Galileo to Newton 1630-1720*. New York: Harper, 1963. (T, L)

A continuation of the account presented by the author's wife, Marie Boas Hall, in *The Scientific Renaissance* (see above).

________. The Scientific Revolution 1500-1800. London: Longmans, 1954; Boston: Beacon Press, 1954. (P) (18-21)

An exceptionally well-written book in which the author deals with the development of scientific ideas and methods during the critical period from the beginning of the 16th century to the end of the 18th century. (M.G. Ebison)

Harré, R. Matter and Method. London: Macmillan, 1964. (18-21)

The author combines a summary of the general conceptual systems underlying the structure of scientific thought in the 17th and 18th centuries with an examination of the metaphysical assumptions made at that time in the development of the corpuscular theory of matter.

Kargon, Robert. Atomism in England from Hariot to Newton. Oxford and New York: Oxford University Press, 1966. (L)

Koyré, Alexandre. Metaphysics and Measurement: Essays in Scientific Revolution. Cambridge, Mass.: Harvard University Press; London: Chapman & Hall, 1968. 165pp. (T, L)

Contents: Galileo and the scientific revolution of the 17th century. Galileo and Plato. Galileo's treatise De motu gravium: the use and abuse of imaginary experiment. An experiment in measurement. Gassendi and science in his time. Pascal savant.

________. Newtonian Studies. Cambridge, Mass.: Harvard University Press, 1965. 288pp. Phoenix Books (University of Chicago Press). (P) French edition: Études Newtoniennes. Paris: Gallimard, 1968. (T, L)

I. The significance of the Newtonian synthesis.
II. Concept and experience in Newton's scientific thought.
III. Newton and Descartes.
IV. Newton, Galileo, and Plato.
V. An unpublished letter of Robert Hooke to Isaac Newton.
VI. Newton's "Regulae philosophandi."
VII. Attraction, Newton, and Cotes.

Toulmin, Stephen and Goodfield, June. The Architecture of Matter. New York: Harper & Row, 1962; Harper Torchbook Paperback, 1966. (18-21)

Scott, J.F. The Scientific Work of René Descartes. London: Taylor & Francis, 1952.

A book which places the main mathematical and physical discoveries of Descartes in an accessible form for English readers. (M.G. Ebison)

Wolf, A. A History of Science, Technology, and Philosophy in the 16th and 17th Centuries. 2nd ed.; 2 vols. New York: Macmillan, 1950; Magnolia, Mass.: Peter Smith Pub.; Harper Torchbook. (P) (T, L)

_______. A History of Science, Technology, and Philosophy in the 18th Century. London: Allen & Unwin, 1938; 2nd ed.; rev. by D. McKie, 1952; Harper Torchbook, 1961. (P)

Excellent reference work for the teacher. Scientific development considered by subjects. (B. Gee)

5. PHYSICAL SCIENCE IN THE 19th AND 20th CENTURIES

d'Abro, A. The Rise of the New Physics, Its Mathematical and Physical Theories (formerly entitled Decline of Mechanism). Princeton: D. Van Nostrand, 1939; 2nd ed., New York: Dover Publications, 1951. (P) (T)

> Volume I is a survey of 19th-century theoretical physics; Volume II is entirely on quantum theory.

Čapek, Milič. Philosophical Impact of Contemporary Physics. Princeton: D. Van Nostrand Co., 1961. 414pp. (P) (T)

> Contrary to the implication of its title, this book is concerned with philosophical problems in physics rather than with the impact of physics on philosophy. It provides a good survey of the basic attitudes imbedded in classical physics and their disintegration as a result of the transition to 20th century theories.

Crombie, A.C., et al. Turning Points in Physics. Amsterdam: North-Holland Pub. Co., 1959; Harper Torchbook paperback, 1961. German trans. by R. Seebass and K. Müller. Wendepunkte in der Physik. Braunschweig: Vieweg, 1963. (20-24)

> Lectures given at Oxford University in 1958 by R.J. Blin-Stoyle, D. ter Haar, K. Mendelssohn, G. Temple, F. Waismann, and D.H. Wilkinson.

Einstein, Albert, and Infeld, Leopold. The Evolution of Physics. New York: Simon & Schuster, 1938; Paperback, 1961. German trans. Die Evolution der Physik, Von Newton bis zur Quantentheorie. Hamburg: Rohwolt, 1959. (18-20)

> While not reliable as a history of modern physics, this account of the development of relativity and quantum theory seems to be attractive to students.

Günther, Siegmund. Geschichte der anorganischen Naturwissenschaften im neunzehnten Jahrhundert. Berlin: Georg Bondi, 1901. (20-25, T)

> Ein Spezialwerk über das 19. Jahrhundert, historisch kompetent für seine Zeit und interessant wegen der Nähe des Verfassers zu jener Zeit. (H. Kangro)

Jaki, S.L. The Relevance of Physics. Chicago: University Press, 1966. (L)

The author, who is a Roman Catholic priest with doctorates in both physics and theology, and who has been a Research Professor in the History and Philosophy of Science, has written a brilliant book in which he seeks to evaluate the scope and limitations of physics. (M.G. Ebison)

Kuznecov, B.G. Von Galilei bis Einstein, Entwicklung der physikalischen Ideen, Übersetzung der russischen Ausgabe (1966). Basel: C.F. Winter: Braunschweig: Vieweg, 1970.

Merz, John Theodore. A History of European Thought in the Nineteenth Century. Vols. I and II. Edinburgh: Blackwood, 1904, 1912; New York: Dover Publications. (P) (L)

Although in many respects out of date, this book provides a wealth of information about 19th-century science, from a comprehensive viewpoint that has not been attempted in any major work since.

Sharlin, Harold I. The Convergent Century: The Unification of Science in the 19th Century. New York: Abelard-Schumann, 1966. (20-24)

Shopl'skii, Eduard V. Ocherkii po istorii Razvitiia sovetskoi Fiziki, 1917-1967 [Essays on the History of the Development of Soviet Physics, 1917-1967].

Thomson, George Paget. The Inspiration of Science. London & New York: Oxford University Press, 1961; Garden City, N.Y.: Doubleday Anchor Books, 1968. (P) (17-20)

A series of essays on physicists of the late 19th and early 20th centuries, with descriptions of their discoveries.

Wilson, William. A Hundred Years of Physics. London: Duckworth, 1950; New York: Humanities Press; Chester Springs, Pa.: Dufour.

6. MONOGRAPHS AND SOURCEBOOKS ON TOPICS IN THE HISTORY OF PHYSICS: MECHANICS

Centore, F.F. Robert Hooke's Contributions to Mechanics, A Study in 17th Century Natural Philosophy. The Hague: Martinus Nijhoff, 1970. (18-21)

A very well-written study of Hooke's place in the history of physics in terms of his accomplishments in the area of mechanics. (M.G. Ebison)

Clagett, Marshall. The Science of Mechanics in the Middle Ages. Madison, Wisc.: University Press, 1959. O/P (L)

Includes many (otherwise unavailable) texts and translations.

Clavelin, Maurice. La Philosophie naturelle de Galilee. Essai sur les origines et la formation de la mecanique classique. Paris: Colin, 1968. 505pp.

Drake, Stillman, and Drabkin, I.E. Mechanics in Sixteenth-Century Italy; Selections from Tartaglia, Benedetti, Guido Ubaldo & Galileo. Madison, Milwaukee & London: University of Wisconsin Press, 1969. O/P in U.S.A. (L)

Sources, annotated and translated.

Dugas, René. Histoire de la Mécanique. Neuchâtel: Editions du Griffon, 1950. English trans., A History of Mechanics. New York: Central Book Co., 1955. O/P (L)

_______. La Mécanique au 17e Siècle. Paris: Dunod, 1954. English trans., Mechanics in the Seventeenth Century. New York: Central Book Co., 1958. O/P (L)

Duhem, Pierre. L'Évolution de la Mécanique. Paris: Joania, 1903.

Hall, A. Rupert. Ballistics in the Seventeenth Century. Cambridge: University Press, 1952. (L)

The author describes the stages of evolution of a mathematical theory of the flight of projectiles. (M.G. Ebison)

Hart, Ivor B. The Mechanical Investigations of Leonardo da Vinci. London: Chapman and Hall, 1925; 2nd ed.; University of California Press, 1963. (P)

Hiebert, Erwin N. *Historical Roots of the Principle of Conservation of Energy*. Madison, Wisc.: State Historical Society of Wisconsin, 1962. O/P (L)

Analysis of relevant works in mechanics up to 1750.

Herivel, John. *The Background to Newton's Principia, A Study of Newton's Dynamical Researches in the Years 1664-84*. Oxford: Clarendon Press, 1965. 337pp. (L)

1. The main line of development of Newton's dynamical thought.
2. The influence of Galileo and Descartes on Newton's dynamics.
3. Newton's concept of *conatus*.
4. Tests of the Law of Gravitation against the moon's motion.
5. The motion of extended bodies.
6. Order of composition and dating of manuscripts. (The last 2/3 of the book consists of manuscripts.)

Hooykaas, R. *Das Verhältnis von Physik und Mechanik in historischer Hinsicht*. Wiesbaden: Steiner, 1963. 22pp.

Beitrage zur Geschichte der Wissenschaft und der Technik Veröffentlichung der Deutschen Gesellschaft für Geschichte der Medizin, Naturwissenschaft und Technik.

Koyré, Alexandré. *Études Galileennes*. Paris: Hermann, 1966. (T)

Reprint of three studies published in the 1930s: *A L'Aube de la Science Classique; La loi de la chute des corps——Descartes et Galilée; Galilée et loi d'inertie*.

________. *A Documentary History of the Problem of Fall from Kepler to Newton. De Motu Gravium Naturaliter Cadentium in Hypothesi Terrae Motae*. Philadelphia: American Philosophical Society, 1955, 1963; originally published as Volume 45, Part 4 of its Transactions. (P) (L)

Mach, Ernst. *Die Mechanik in ihrer Entwicklung, historisch-kritisch dargestellt*. Leipzig: Brockhaus, 1883; 4th ed. 1901. English trans. by T.J. McCormack, *The Science of Mechanics, A Critical and Historical Account of Its Development*. 6th ed., LaSalle, Ill.: The Open Court Pub. Co., 1960. (P) (T, L)

This book is itself of sufficient historical importance (for example, in undermining the concepts of absolute space and time) to compensate for its deficiencies as a history.

Maier, Anneliese. *Zwei Grundprobleme der scholastischen Naturphilosophie, Das Problem der intensiven Grösse & Die Impetustheorie*. 2nd ed. Roma: Edizioni di Storia e letteratura. 1951. (L)

________. *Die Vorläufer Galileis im XIV Jahrhundert*. Roma: Edizioni di Storia e letteratura, 1949. 307pp. (L)

Pogrebissky, I.B. *Ot Lagranzha k Eenshtyeenoo, Klassicheskaya Mekhanika XIX veka* [*From Lagrange to Einstein, Classical Mechanics in the 19th Century*]. Moscow: Izdatelstva "Nauka," 1966. (L)

Rouse, Hunter, and Ince, Simon. *History of Hydraulics*. Iowa Institute of Hydraulic Research, 1957; New York: Dover Publications, 1963. (P) (L)

Seeger, Raymond J. *Men of Physics: Galileo Galilei, His Life and Works*. Oxford and New York: Pergamon Press, 1966. (P) (19-22, T)

Includes extracts from Galileo's writings.

Theobald, D.W. *The Concept of Energy*. London: E. & F.N. Spon; New York: Barnes & Noble, 1966.

Timoshenko, Stephen. *History of Strength of Materials, with a Brief Account of the History of Theory of Elasticity and Theory of Structures*. New York: McGraw-Hill, 1953.

Truesdell, C. *Rational Fluid Mechanics, 1687-1765*. L. Euleri Opera Omnia (2) Vol. 12, IX-CXXV. Zürich: Orell Füssli, 1954. (L)

________. I. *The First Three Sections of Euler's Treatise on Fluid Mechanics (1760)*; II. *The Theory of Aerial Sound (1687-1788)*; III. *Rational Fluid Mechanics (1765-1788)*. L. Euleri Opera Omnia (2) Vol. 13, VII-CXVIII. Zürich: Orell Füssli, 1956. (L)

________. *The Rational Mechanics of Flexible or Elastic Bodies, 1638-1788*. L. Euleri Opera Omnia (2) Vol. 112. Zürich: Orell Füssli, 1960. (L)

________. *Essays in the History of Mechanics*. New York: Springer-Verlag, 1968. 384pp. (T, L)

I. The mechanics of Leonardo da Vinci.

II. A program toward rediscovering the rational mechanics of the Age of Reason.
III. Reactions of late Baroque mechanics to success, conjecture, error, and failure in Newton's Principia.
IV. The creation and unfolding of the concept of stress.
V. Whence the law of moment of momentum?
VI. Early kinetic theories of gases.
VII. Reactions of the history of mechanics upon modern research.
VIII. Recent advances in rational mechanics. (1956)

7. MONOGRAPHS AND SOURCE BOOKS ON TOPICS IN THE HISTORY OF PHYSICS: OPTICS

Ames, J.S. Prismatic and Diffraction Spectra, Memoirs by Joseph von Fraunhofer. New York: American Book Co., 1900. (20-24, T, L)

Bradbury, S. The Evolution of the Microscope. Oxford and New York: Pergamon Press, 1967. (20-24)

A fascinating account of the development of this most important of scientific instruments. The author provides a valuable source of reference which will last for many years. The text is enlivened by a large number of illustrations. (M.G. Ebison)

Bradbury, S., and Turner, G. L'E., eds. Historical Aspects of Microscopy. Cambridge: Heffer & Sons, Ltd., 1967. (20-24)

A collection of six essays on the history of the scientific study of vision and the development of the microscope. The volume was published for the Royal Microscopical Society to mark the centenary of the granting of the Royal Charter to the Society. (M.G. Ebison)

Crew, Henry. *The Wave Theory of Light, Memoirs by Huygens, Young, and Fresnel*. New York: American Book Co., 1900. (20-24, T, L)

Harvey, E.N. *A History of Bioluminescence, from the Earliest Times until 1900*. Philadelphia: American Philosophical Society, 1957. (L)

Hoppe, Edmund. *Geschichte der Optik*. Leipzig: J.J. Weber, 1926; Wiesbaden: M. Sändig, 1967.

King, H.C. *The History of the Telescope*. London: Charles Griffin and Co., Ltd., 1955. (20-24)

A well-written comprehensive history of the telescope from its invention to the construction of the great 200-inch Hale reflector on Palomar Mountain in California. (M.G. Ebison)

Klemm, F. *Die Geschichte der Emissionstheorie des Lichts*. Weimar: R. Borkmann, 1932.

Dissertation: mit guter historischer Sachkenntnis und vorzüglicher Literaturkenntnis gegebene, zuverlässige Darstellung. (H. Kangro)

MacAdam, David L., ed. *Sources of Color Science*. Cambridge, Mass.: The M.I.T. Press, 1970.

Mach, Ernst. *Die Prinzipien der physikalischen Optik, historisch und erkenntnis-psychologisch entwickelt*. Leipzig: J.A. Barth, 1921. English trans. by J.S. Anderson and A.F.A. Young, *The Principles of Physical Optics, An Historical and Philosophical Treatment*. London: Methuen, 1926; New York: Dover Pubs. (P)

McGucken, William. *Nineteenth-Century Spectroscopy*. Baltimore: The Johns Hopkins Press, 1970.

A description of the development of the understanding of spectra between 1802 and 1897 together with an analysis of the effect that this understanding had upon ideas concerning the atom. (M.G. Ebison)

Partington, J.R. *An Advanced Treatise on Physical Chemistry*. Vol. 4. *Physico-Chemical Optics*. London, New York & Toronto: Longmans, Green and Co., 1953. New York: Wiley. (L)

A detailed technical account, arranged by topics, with extensive references to original sources.

Ronchi, Vasco. The Nature of Light. Trans. from 2nd Italian ed. by V. Barocas. Cambridge, Mass.: Harvard University Press, 1970. 200pp.

________. Optics, The Science of Vision. Trans. by E. Rosen. New York: New York University Press, 1957. O/P (L)

Includes much of the material of his Histoire de la Lumiere.

________. Storia della luce. 2nd ed. Bologna: Edizione Zanichelli. 288pp.

Sabra, A.I. Theories of Light from Descartes to Newton. London: Oldbourne, 1967, 363pp; New York: American Elsevier. (T)

A detailed study of the major 17th-century works, including Fermat and Huygens as well as Descartes and Newton.

Sanders, John H. The Velocity of Light. Oxford & New York: Pergammon Press, 1965. (P) (21-24)

Includes reprints of original sources.

Tolansky, Samuel. Revolution in Optics. Harmondsworth: Penguin, 1968. (L)

Includes a brief historical survey of waves, particles, quantum theory and the expanding universe; the main part of the book is devoted to advances in instrumentation in the 20th century.

Whittaker, E.T. A History of the Theories of Aether and Electricity. Rev. ed. in two vols.; London: Thomas Nelson and Sons, Ltd., and New York: Humanities Press, 1951; Harper Torchbook ed. O/P

Vol. I: The classical theories. Optics, electricity and magnetisim in the 17th, 18th and 19th centuries; aether models; the electron and radiation theory. Vol. II: The modern theories. Rutherford, Poincaré, Lorentz, relativity, spectroscopy, quantum theory.

Wilde, E. Geschichte der Optik. Berlin, 1838; Wiesbaden: M. Sändig, 1968.

8. MONOGRAPHS AND SOURCE BOOKS ON TOPICS IN THE HISTORY OF PHYSICS: ELECTRICITY AND MAGNETISM

Ames, J.S. The Discovery of Induced Electric Currents. Vol. 1, Memoirs by Joseph Henry. Vol. 2, Memoirs by Michael Faraday. New York: American Book Co., 1900.

Daujat, J. Origines et Formation de la Theorie des Phenomenes magnetiques et electriques. Paris, 1945.

Fierz, Markus. Die Entwicklung der Elektrizitätslehre als Beispiel der physikalischen Theorienbildung. Basel: Buchdr. zum Basler Berichthaus, 1951.

Hoppe, E. Geschichte der Elektricität. Leipzig: J.A. Barth, 1884.

Lewis, E.P. The Effects of a Magnetic Field on Radiation, Memoirs by Faraday, Kerr, and Zeeman. New York: American Book Co., 1900.

Ostwald, W. Elektrochemie, ihre Geschichte und ihre Lehre. Leipzig: Veit & Comp., 1896.

> Hinter dem Titel dieses umfangreichen Werkes (ca. 1150 Seiten) des historisch interessierten Verfassers verbirgt sich eine Menge Material über die Geschichte der Elektrizität in flüssigen Leitern. Es is noch heute in der Physikgeschichte unentbehrlich. (H. Kangro)

Priestley, Joseph. The History and Present State of Electricity, with Original Experiments. Reprinted from 3rd ed. London, 1755; New York: Johnson Reprint Corp., 1966.

Roller, Duane, and Roller, Duane H.D. The Development of the Concept of Electric Charge: Electricity from the Greeks to Coulomb. (In Harvard Case Histories in Experimental Science, ed. J.B. Conant; also available as a separate pamphlet.) Cambridge, Mass.: Harvard University Press, 1950. (18-22)

> Analysis of the works of William Gilbert, Stephen Gray, Charles DuFay, Benjamin Franklin, Joseph Priestley, and Coulomb.

Tricker, R.A.R. Early Electrodynamics, The First Law of Circulation. Oxford & New York: Pergamon Press, 1965. (20-24)

> Includes extracts from writings of Ampere and others.

_______. The Contributions of Faraday & Maxwell to Electrical Science. Oxford & New York: Pergamon Press, 1966. (20-24)

Includes extracts from their writings.

Whittaker, E.T. A History of the Theories of Aether and Electricity. 2nd ed.; London: Thomas Nelson & Sons, Ltd., & New York: Humanities Press, 1961; Harper Torchbook paperback, O/P (T)

Williams, L. Pearce. The Origins of Field Theory. New York: Random House, 1966. (P) (20-24)

Yajima, S. Denziriron no Hattensi [Development of the Electromagnetic Theory]. Tokyo: 1947.

_______. Denzikigaku-si [A History of Electromagnetism]. Tokyo: 1950.

9. MONOGRAPHS AND SOURCE BOOKS ON TOPICS IN THE HISTORY OF PHYSICS: HEAT, GASES, KINETIC THEORY

Ames, J.S. The Free Expansion of Gases. Memoirs by Gay-Lussac, Joule, and Joule and Thomson. New York: Harper, 1898.

Bachelard, G. Étude sur l'Evolution d'un Probleme de Physique: La Propagation thermique dans les Solides. Paris: Librairie philosophique J. Vrin, 1927, 1946. (T)

A major work of Bachelard; mainly philosophical. (G. Boutry)

Barus, Carl. The Laws of Gases. Memoirs by Robert Boyle and E.G. Amagat. New York: Harper, 1899.

Brace, D.B. The Laws of Radiation and Absorption. Memoirs by Prevost, Stewart, Kirchhoff, and Kirchhoff and Bunsen. New York: American Book Co., 1901.

Brush, Stephen G. Kinetic Theory. Vol. 1, The Nature of Gases and of Heat (1965). Vol. 2. Irreversible Processes (1966). Vol. 3, The Chapman-Enskog Solution of the Transport Equations for Moderately Dense Gases (1972). Oxford & New York: Pergamon Press. (P) (20-24)

Each volume includes reprints and translations of sources.

Conant, James Bryant. Robert Boyle's Experiments in Pneumatics. (In Harvard Case Histories in Experimental Science, ed. J.B. Conant; also available as a separate pamphlet.) Cambridge, Mass.: Harvard University Press, 1950. (19-22)

Includes extensive quotations from Boyle's writings.

Dugas, René. La Theorie physique au Sens de Boltzmann et ses Prolongements modernes. Neuchatel-Suisse: Editions de Griffon, 1959. 308pp. (T)

An account of Boltzmann's theories of gases, and his debates with Loschmidt and Zermelo on the H-theorem, reversibility paradox, and recurrence paradox; relevant discoveries in the 20th century (quantum theory, Brownian movement) are also touched on.

Fox, Robert. The Caloric Theory of Gases from Lavoisier to Regnault. Oxford: Clarendon Press, 1971. (T, L)

A fully documented analytical account of theories and experiments concerning the nature of heat, from about 1780 to 1850.

Mach, Ernst. Die Principien der Wärmelehre, historisch-kritisch entwickelt. Leipzig: J.A. Barth, 1896. (L)

McKie, Douglas, and Heathcote, N.H. deV. The Discovery of Specific and Latent Heats. London: Arnold, 1935. O/P

Mendelssohn, K. The Quest for Absolute Zero. London: Weidenfeld & Nicolson, 1966; New York: McGraw-Hill, 1966. (P) (18-21)

An absorbing account. (M.G. Ebison)

Mendoza, E. Reflections on the Motive Power of Fire, by Sadi Carnot, and other Papers on the Second Law of Thermodynamics by E. Clapeyron and R. Clausius. New York: Dover, 1960. (Based in part on the 1899 edition of W.F. Magie.)

Meyer, Kirstine. Die Entwicklung des Temperaturbegriffs im Laufe der Zeiten sowie dessen Zusammenhang mit den wechselnden Vorstellungen über die Natur der Wärme. Braunschweig: Vieweg, 1913.

Vorzügliche Darstellung. (H. Kangro)

Middleton, W.E.K. The History of the Barometer. Baltimore: The Johns Hopkins Press, 1964. (20-24)

This is a lively and authoritative account of the development of the barometer from its invention to the present time. (M.G. Ebison)

________. A History of the Thermometer and Its Use in Meteorology. Baltimore: The Johns Hopkins Press, 1967.

Mott-Smith, Morton. The Story of Energy. New York: Appleton-Century, 1934; reprinted as The Concept of Energy Simply Explained. New York: Dover Pubs., 1964. (P)

Randall, Wyatt W. The Expansion of Gases by Heat. New York: American Book Co., 1902. Memoirs by Dalton, Gay-Lussac, Biot, Regnault, and Chappuis.

Roller, Duane. The Early Development of the Concepts of Temperature and Heat. The Rise and Decline of the Caloric Theory. (In Harvard Case Histories in Experimental Science, ed. J.B. Conant; also available as a separate pamphlet.) Cambridge, Mass.: Harvard University Press, 1950. (18-22)

Includes extensive quotations from the writings of Joseph Black, Rumford, and Davy; bibliography, and questions for the student.

Takabayasi, T. Netugaku-si [A History of the Science of Heat]. Kyoto, 1948.

Theobald, D.W. The Concept of Energy. London: E. & F.N. Spon; and New York: Barnes & Noble, 1966.

Watanabe, M. Bunkasi ni okeru Kindai Kagaku [Science in the History of Modern Culture]. Tokyo: Miraisha, 1963.

A collection of articles, including five papers on the development of the science of heat in England, and six papers concerning the role of history of science in science education. English summaries are appended to each article.

10. MONOGRAPHS AND SOURCE BOOKS ON TOPICS IN THE HISTORY OF PHYSICS: QUANTUM THEORY

Amano, K. Netuhukusyaron to Ryosiron no Kigen [Theory of Heat Radiation and the Origin of Quantum Theory]. Tokyo, 1943.

> This book contains a historical introduction by the late Kiyosi Amano as well as the papers by W. Wien, Lord Rayleigh, and M. Planck. Amano's introduction is in fact an excellent article on the history of the experimental and theoretical researches which led to Planck's discovery of the energy quantum. It is distinguished from many other articles on the birth of quantum theory by its analysis of the experimentation and of the technological background of researches on heat radiation. (T. Hirosige)

________. Ryosirikigaku-si [A History of Quantum Mechanics]. Kyoto, 1948.

Cline, Barbara Lovett. Men Who Made A New Physics——Physicists and the Quantum Theory. Crowell: 1965. The New American Library (Signet paperback), 1969. (15-19)

> Mainly biographical, with elementary explanations of some of the physical concepts. Recommended for students. (S. Brush)

Gamow, George. Thirty Years That Shook Physics, The Story of Quantum Theory. Garden City, N.Y.: Doubleday & Co., 1966. (P) (19-23)

> An anecdotal account based on the author's own recollections.

Guillemin, Victor. The Story of Quantum Mechanics. New York: Scribner, 1968. (P)

Haar, D. ter. The Old Quantum Theory. Oxford & New York: Pergamon Press, 1967. (P) (21-24)

> Includes reprints and translations of sources.

Hermann, Armin, ed. Die Quantentheorie der spezifischen Wärme. München: Battenberg, 1967.

> Includes facsimiles of papers by Einstein, Debye, Born and von Kármán, with an introductory essay by the editor.

______. Frühgeschichte der Quantentheorie 1899-1913. Baden: Mosbach, 1969. English trans. to be published by the MIT Press, Cambridge, Mass.

Hindmarsh, W.R. Atomic Spectra. Oxford & New York: Pergamon Press, 1967. (21-24)

An exposition of the subject is followed by extensive extracts from original sources.

Hoffman, Banesh. The Strange Story of the Quantum. New York: Harper, 1947; 2nd ed.; Dover, 1959. 285pp. (P) (17-21)

Hund, F. Geschichte der Quantentheorie. Mannheim: Bibliographisches Institut, 1967.

Darstellung der Theorie vom Standpunkte des Physikers, nicht so sehr des Historikers. (H. Kangro)

Jammer, Max. The Conceptual Development of Quantum Mechanics. New York: McGraw-Hill, 1966. 399pp. (T)

Planck's quantum theory; line spectra; wave mechanics and matrix mechanics; statistical transformation theory; philosophical aspects.

Kangro, Hans. Vorgeschichte des Planckschen Strahlungsgesetzes. Messungen und Theorie der spektralen Energieverteilung bis zur Begrundung der Quantenhypothese. Wiesbaden: F. Steiner Verlag GMBH, 1970. (P) (T, L)

A detailed account of experimental and theoretical studies of black body radiation from 1880 to 1901.

Ludwig, Gunther. Wave Mechanics. Oxford & New York: Pergamon Press, 1968. (P) (22-24)

Exposition of the subject with extracts and translations of original sources.

Meyer-Abich, Klaus Michel. Korrespondenz, Individualität und Komplementarität, eine Studie zur Geistesgeschichte der Quantentheorie in den Beiträgen Niels Bohrs. Wiesbaden: F. Steiner, 1965. (T)

Philosophische, weniger physikalische Darstellung. (H. Kangro)

Peterson, Aage. Quantum Physics and the Philosophical Tradition. Cambridge, Mass.: The M.I.T. Press, 1968. 212pp.

Schwinger, Julian. Selected Papers on Quantum Electrodynamics. New York: Dover Publications, 1958. (P) (T, L)

Aside from a brief preface this consists entirely of reprints of papers.

Taketani, M. Ryosirikigaku no Keisei to Ronri [The Formation and Logic of Quantum Mechanics]. Vol. 1. Tokyo, 1948.

van der Waerden, B.L. Sources of Quantum Mechanics. Amsterdam: North-Holland Pub. Co., 1967; New York: Dover Publications. (P) (22-24)

Historical introduction, and reprints or translations of the papers leading up to and including the formulation of matrix mechanics.

11. MONOGRAPHS AND SOURCE BOOKS ON TOPICS IN THE HISTORY OF PHYSICS: RELATIVITY AND COSMOLOGY

d'Abro, A. The Evolution of Scientific Thought from Newton to Einstein. Boni & Liveright, 1927; 2nd ed.; New York: Dover Publications, 1950. (P) London: Constable & Co.; Magnolia, Mass.: Peter Smith, Pub. (20-24)

Charon, Jean. Cosmology. Trans. from French by P. Moore. London: Weidenfeld & Nicolson, 1970. Also published under title Theories of the Universe. (P) (17-20)

A good elementary account of the development of ideas about the universe.

Dickson, F.P. The Bowl of Night; The Physical Universe and Scientific Thought. Eindhoven: N.V. Philips' Gloeilampenfabrieken, 1968; Cambridge, Mass.: M.I.T. Press. (P) (20-24, T)

Frankfurt, Usher I. Spetsial'naia i obshchaia teoriia otnositel'nosti; istoricheskie ocherki [Special and General Theory of Relativity; Historical Outline]. Moscow: Nauka, 1968.

Graves, J.C. The Conceptual Foundations of Contemporary Relativity Theory. Cambridge, Mass.: The M.I.T. Press, 1971. (T, L)

A fascinating historical and philosophical examination of the physical principles of relativity. Thoroughly to be recommended. (M.G. Ebison)

Holton, Gerald, et al. Special Relativity Theory, Selected Reprints. New York: American Institute of Physics, 1963. (P) (T, L)

Includes Holton's "Resource Letter" (1962) and articles by Bondi, Dingle, Crawford, Terrell, Weisskopf, Frisch, and others, mostly on the "clock paradox."

Kilmister, C.W. Special Theory of Relativity. Oxford: Pergamon Press, 1970. (20-24)

The first part of this book deals with a historical introduction, Einstein's contribution, the consequences of the Lorentz transformation and applications of special relativity in quantum theory. This part is followed by reprints and translations of original sources. (M.G. Ebison)

North, J.D. The Measure of the Universe, A History of Modern Cosmology. Oxford: Clarendon Press, 1965. (T)

19th-century astronomy: The nebulae, cosmological difficulties with the Newtonian theory of gravitation. Einstein's theory. The expanding universe. Theories of Birkhoff and Whitehead, Milne. Steady-state theories. Philosophical issues. Recommended for teachers of advanced courses in relativity and cosmological theories. For students, Dickson's text The Bowl of Night would be more suitable.

Williams, L. Pearce. Relativity Theory: Its Origins and Impact on Modern Thought. New York: Wiley, 1969. (P) (20-24)

Includes extracts from writings of physicists, historians of science, and popular writers.

12. MONOGRAPHS AND SOURCE BOOKS ON TOPICS IN THE HISTORY OF PHYSICS: ATOMIC AND NUCLEAR PHYSICS (19th and 20th Centuries). [See #13 for books on the earlier history of atomism.]

Anderson, David L. The Discovery of the Electron; The Development of the Atomic Concept of Electricity. Princeton, N.J.: D. Van Nostrand Co., 1964. A Momentum Book. (P) (19-23)

Bacon, G.E. X-Ray and Neutron Diffraction. Oxford: Pergamon Press, 1966. (P) (22-24)

> A pedagogical introduction to the subject, followed by extensive extracts from sources.

Bagge, E., Diebner, K., and Jay, K. Von der Uranspaltung bis Calder Hall. Hamburg: Rowohlt, 1957. (20-26)

> Diese Schrift enthält den von E. Bagge u.K. Diebner selbst miterlebten historischen Hintergrund der Ereignisse in Deutschland vor und während des Krieges sowie nach dem Kriege in Frankreich und England, ferner die von allen drei Verfassern erlebte Entwicklung nach dem Kriege. (H. Kangro)

Barker, George F. Röntgen Rays; Memoirs by Röntgen, Stokes, and J.J. Thomson. New York: Harper, 1899.

Beyer, Robert T. Foundations of Nuclear Physics. New York: Dover Pubs., 1949. (P) (22-4, L)

> Reprints of 13 papers and a 120-page bibliography.

Boorse, H.A., and Motz, Lloyd. The World of the Atom. New York: Basic Books, 1966.

Brink, D.M. Nuclear Forces. Oxford & New York: Pergamon Press, 1965. (P) (22-24)

> Includes reprints and translations of original sources.

Brock, W.H., ed. The Atomic Debates. Leicester, England: Leicester University Press, 1967. 186pp.; New York: Humanities. (20-24)

> An absorbing account of one 19th-century chemist's serious alternative to Dalton's atomic theory in the form of operationalism which anticipated Bridgman's use of the term by 60 years. (M.G. Ebison)

Cantore, E. Atomic Order. Cambridge, Mass.: M.I.T. Press, 1969. (L, T)

This book is in two parts. The first is devoted to the problem of atomic order as discovered and verified by physicists, from Dalton's work to contemporary quantum mechanics. The second part of the book is more philosophical in nature, analyzing the presuppositions, guiding principles and implications that made the development of atomic physics possible. (M.G. Ebison)

Conn, G.K., and Turner, H.D. The Evolution of the Nuclear Atom. New York: American Elsevier, 1966.

Friedman, F.L., and Sartori, L. The Classical Atom. Reading, Mass.: Addison-Wesley, 1965. (18-21)

This book concentrates on the identification of subatomic particles and on atomic models culminating in the nuclear atom early in the 20th century. (M.G. Ebison)

Frisch, O.R., et al., eds. Beiträge zur Physik und Chemie des 20. Jahrhunderts; Lise Meitner, Otto Hahn, Max von Laue zum 80. Geburtstag. Braunschweig: Vieweg, 1959.

Articles on the history of atomic and nuclear physics.

Graetzer, H.G., and Anderson, D.L. The Discovery of Nuclear Fission, A Documentary History. New York: Van Nostrand Reinhold, 1971.

Hanson, N.R. The Concept of the Positron. Cambridge: University Press, 1963. (L, T)

In this book the author examines the historical foundations and conceptual structures underlying the discovery of the first 'antiparticle' to be recognized and identified. (M.G. Ebison)

Knight, D.M. Atoms and Elements. London: Hutchinson, 1967. (18-21)

A very well-written account of the reception of Dalton's atomic theory in the 19th century.

Livingston, M.S. Particle Accelerators: A Brief History. Cambridge, Mass.: Harvard University Press, 1969. (L)

A readable account of the development of particle accelerators from the electrostatic generators to the 200 GeV accelerator. (M.G. Ebison)

McCarthy, I.E. Nuclear Reactions. Oxford & New York: Pergamon Press, 1970. (P) (22-24)

Includes reprints of original papers.

Romer, Alfred. The Discovery of Radioactivity and Transmutation. New York: Dover Publications, 1964. (P) (22-24)

Reprints and translations of sources, with notes and commentary.

________. Radiochemistry and the Discovery of Isotopes. New York: Dover Publications, 1970. (P) (22-24)

Reprints and translations of sources, with commentary and an introductory historical essay.

Schonland, Basil. The Atomists 1804/1933. Oxford and New York: Clarendon Press, 1968. (18-21)

Strachan, Charles. The Theory of Beta-Decay. Oxford & New York: Pergamon Press, 1969. (P) (22-24)

Includes reprints of original sources.

Trigg, G.L. Crucial Experiments in Modern Physics. New York: Van Nostrand Reinhold, 1971 (P)

13. HISTORIES OF INDIVIDUAL CONCEPTS

Hesse, Mary. Forces and Fields, the Concept of Action at a Distance in the History of Physics. London: Thomas Nelson and Sons Ltd., 1961; Totowa, N.J.: Littlefield, Adams. (P) (T)

Jammer, Max. Concepts of Force, A Study in the Foundations of Dynamics. Cambridge, Mass.: Harvard University Press, 1957. 269pp. New York: Harper Torchbooks. (P) (T)

_______. Concepts of Mass in Classical and Modern Physics. Cambridge, Mass.: Harvard University Press, 1961. 230pp. New York: Harper Torchbooks, 1964. (T)

_______. Concepts of Space, the History of the Theories of Space in Physics. Cambridge, Mass.: Harvard University Press, 1954; 2nd ed.; New York: Harper Torchbooks, 1960. (P) German trans. by P. Wilpert, Das Problem des Raumes, die Entwicklung der Raumtheorie. Darmstadt: Wissenschaftliche Buchgesellschaft, 1960. (T)

> Diese Übersetzung enthält eine grössere Anzahl von Fehlern auf Grund der fehlenden Sachkenntnis des Übersetzers. (H. Kangro)

Lanczos, Cornelius. Space through the Ages, the Evolution of Geometrical Ideas from Pythagoras to Hilbert and Einstein. New York: Academic Press, 1970. (T)

Lasswitz, K. Geschichte der Atomistik vom Mittelalter bis Newton. Hamburg & Leipzig: 1890; Hildesheim: G. Olms, 1963. (L)

> Immer noch ein vorzüglish brauchbares Werk. (H. Kangro)

McKenzie, A.S. The Laws of Gravitation, Memoirs by Newton, Bouguer, and Cavendish, Together with Abstracts of Other Important Memoirs. New York: American Book Co., 1900.

Scott, Wilson L. The Conflict between Atomism and Conservation Theory, 1644-1860. New York: American Elsevier Pub. Co., 1970.

Van Melsen, Andrew G. From Atomos to Atom, the History of the Concept Atom. Trans. by H.J. Koren from the Dutch edition first published in 1949. Pittsburgh: Duquesne University Press, 1952; Harper Torchbook paperback, 1960. O/P in U.S.A. (T)

Whitrow, G.J. The Natural Philosophy of Time. London: Nelson, 1961.

Whyte, Lancelot L. Essay on Atomism from Democritus to 1600. Middletown, Conn.: Wesleyan University Press, 1961.

Ward, F.A.B. Time Measurement, Historical Review. London: Science Museum, 1970.

This impression is reproduced from the 1961 edition but has an appendix giving addenda and corrigenda.

14. TOPICS IN THE HISTORY OF PHYSICS (MISCELLANEOUS)

Crowe, Michael J. A History of Vector Analysis. Notre Dame: University Press, 1967.

Humphries, Willard C. Anomalies and Scientific Theories. San Francisco: Freeman, Cooper & Co., 1968. O/P in U.S.A. (T)

Includes a case history of meson theory.

Owen, George E. The Universe of the Mind. Baltimore: Johns Hopkins Press, 1971. (P) (20-24)

A comprehensive and readable account of mathematical theories and their relations to the history of physics.

15. BIOGRAPHICAL DICTIONARIES AND COLLECTIONS OF BIOGRAPHICAL ESSAYS

Asimov, Isaac. Asimov's Biographical Encyclopedia of Science and Technology. Garden City, N.Y.: Doubleday, 1964. 662pp. (T)

Debus, Allen G., ed. World Who's Who in Science: A Biographical Dictionary of Notable Scientists from Antiquity to the Present. Chicago: Marquis--Who's Who Inc., 1968. 1855pp. (L)

> About 30,000 entries, including many living scientists.

Gillispie, C.C., ed. Dictionary of Scientific Biography. Vols. I and II. New York: Charles Scribner's Sons, 1970. (L)

> These first two volumes of a series contain articles on scientists from Abailard to Buys-Ballot.

Harré, R., ed. Early Seventeenth Century Scientists. Oxford: Pergamon Press, 1965. (18-21)

> A collection of very readable articles on seven famous scientists of the early 17th century: Gilbert, Francis Bacon, Galileo, Kepler, Harvey, Van Helmont and Descartes.

Hart, Ivor B. The Great Physicists. London: Methuen, 1927; Freeport, N.Y.: Books for Libraries Press, 1970.(18-22)

> Through mid-19th century only.

Heathcote, Niels H. de V. Nobel Prize Winners in Physics 1901-1950. New York: Schuman, 1953. O/P in U.S.A.

Howard, A.V. Dictionary of Scientists. London: Chambers, 1964.

Hutchings, D., ed. Late Seventeenth Century Scientists. Oxford: Pergamon Press, 1969. (18-21)

> Six essays by different authors on the lives and scientific work of important late 17th century scientists: Boyle, Malpighi, Wren, Huygens, Hooke and Newton.

Lenard, Philipp. Great Men of Science, A History of Scientific Progress. Trans. from the 2nd German ed., 1932, by H.S. Hatfield. Freeport, N.Y.: Books for Libraries Press, 1970. (L)

Sketches (5 to 8 pages each) of the lives and discoveries of many physicists. The lengthy discussion of Hasenöhrl and his work on the mass of electromagnetic energy includes several references to Lenard's own theories on the ether but carefully avoids any mention of Einstein.

Leprince-Ringuet, L., ed. Les Inventeurs célèbres. Paris, 1950. ca. 450pp.

Mainly biographical; entries on about 120 scientists, almost all of whom are physicists. Carefully written by more than 60 contributors. Beautifully produced and illustrated; an appendix lists some 1200 names of physicists and engineers not considered in the text, with mention of their main contributions. (G. Boutry)

MacDonald, D.K.C. Faraday, Maxwell, and Kelvin. Garden City, N.Y.: Doubleday Anchor, 1964. (P) (17-20)

Mann, Alfred L, and Vivian, A.C. Famous Physicists. New York: The Day Co., 1961. O/P in U.S.A. (12-16)

Archimedes, Roger Bacon, Galileo, Otto von Guericke, Newton, Franklin, Galvani, Volta, Faraday.

North, John, ed. Mid-Nineteenth-Century Scientists. Oxford: Pergamon Press, 1969. (18-21)

Includes Babbage and Joule.

Olby, R.C., ed. Late Eighteenth Century European Scientists. Oxford: Pergamon Press, 1966. (18-21)

The aim of this book is to give an account of the remarkable progress which was made by European scientists at the close of the 18th century in chemistry, electricity, astronomy, and botany. (M.G. Ebison)

________. Early Nineteenth Century European Scientists. Oxford: Pergamon Press, 1967. (18-21)

Includes biographies of Humphry Davy and Thomas Young.

Williams, Trevor I., ed. A Biographical Dictionary of Scientists. London: Adam & Charles Black, 1969. 592pp. (L)

Erckmann, R., hrsg. Via Regia, Nobelpreisträger auf dem Wege ins Atomzeitalter. München & Wien: W. Andermann, 1958. (T)

Sammlung von wissenschaftlichen Biographien über: Lenard, Röntgen, Laue, Stark, J. Franck und G. Hertz, Braun, Nernst, Wien, Hess, Bothe, Planck, Einstein, Born, Heisenberg, Schrödinger, O. Hahn. (H. Kangro)

16. BIOGRAPHIES AND AUTOBIOGRAPHIES OF PHYSICISTS

[This is only a small and somewhat arbitrary selection; the list could easily be doubled if we tried to include all biographies of all physicists.]

Andre Marie AMPÈRE (1775-1836)

Launay, Louis de. Le grand Ampere. Paris: 1925.

Lewandowski, Maurice. Andre-Marie Ampere, la Science et la Foi. Paris: 1936.

ARCHIMEDES (287?-212 B.C.)

Dijksterhuis, E.J. Archimedes. Groningen: Noordhof, 1938 (in Dutch); English trans. (with additional material) by C. Dikshoorn, Copenhagen: Munksgaard, 1956; New York: Humanities Press.

Gardner, Martin. Archimedes, Mathematician and Inventor. New York: Macmillan, 1965. (10-14)

Benedick, Jeanne. Archimedes and the Door of Science. New York: Watts, 1962. (12-17)

ARISTOTLE (384-322 B.C.)

Jaeger, W. Aristotle. Oxford, 1934; 2nd ed., 1948. (P)

Ross, W.D. Aristotle. 5th ed.; New York: Barnes & Noble, 1955. (P)

Niels BOHR (1885-1962)

Moore, Ruth. Niels Bohr: The Man, His Science, and the World They Changed. New York: Knopf, 1966. German trans., München: P. List, 1970. (20-24)

Rozental, S., ed. Niels Bohr, His Life and Work as Seen by His Friends and Colleagues. Trans. from the Danish edition of 1964; New York: Wiley/Interscience, 1967. (P) (T, L)

Silverberg, Robert E. Niels Bohr, The Man Who Mapped the Atom. Philadelphia: Macrae Smith Co., 1965. (15-18)

Ludwig BOLTZMANN (1844-1906)

Broda, Engelbert. Ludwig Boltzmann. Mensch - Physiker - Philosoph. Wien: Deuticke, 1955; Berlin: VEB. Deutscher Verlag der Wissenschaften, 1957. (20-24)

Marie Sklodowska CURIE (1867-1934)

Curie, Eve. Madame Curie, 1867-1934: A Biography. Trans. by Vincent Sheean. Garden City, N.Y.: Doubleday, 1937; Pocket Books, 1959. (16-20)

Ivimey, Alan. Marie Curie, Pioneer of the Atomic Age. New York: Praeger, 1969.

McKown, Robin. Marie Curie. New York: Putnam, 1959. (10-15)

Thorne, Alice D. The Story of Madame Curie. New York: Grosset, 1959. (P)

Albert EINSTEIN (1879-1955)

Clark, Ronald W. Einstein, The Life and Times. New York: World Pub. Co., 1971. (20-24, T)

Einstein, Albert. Autobiographical Notes (pp. 1-95; German and English on facing pages) in P.A. Schilpp, ed., Albert Einstein Philosopher-Scientist. Evanston, Ill.: The Library of Living Philosophers, 1949; Harper Torchbook paperback edition 1959 (O/P). (20-24)

Frank, Philipp. Einstein, His Life and Times. Trans. from German by G. Rosen. New York: Knopf, 1967. (20-24)

Infeld, Leopold. Albert Einstein. New York: Scribner, 1950. (P)

Jordan, Pascual. Albert Einstein, sein Lebenswerk und die Zukunft der Physik. Frauenfeld & Stuttgart: Verlag Huber, 1969. (T)

Michael FARADAY (1791-1867)

Sootin, Harry. Michael Faraday, from Errand Boy to Master Physicist. New York: Julian Messner, 1954. (12-16)

Williams, L. Pearce. Michael Faraday. New York: Basic Books, 1965. (T, L)

Enrico FERMI (1901-1954)

Segre, Emilio. Enrico Fermi, Physicist. Chicago: University Press, 1970.

Fermi, Laura. Atoms in the Family, My Life with Enrico Fermi. Chicago: University Press, 1954; Phoenix Books. (P) (17-20)

Galileo GALILEI (1564-1642)

Broderick, James. Galileo. New York: Harper, 1965.

Drake, Stillman. Galileo Studies, Personality, Tradition, and Revolution. Ann Arbor: The University of Michigan Press, 1970.

Geymonat, Ludovico. Galileo Galilei, A Biography and Inquiry into His Philosophy of Science. Eng. trans. (from Italian ed. of 1957) by S. Drake, with additional notes and appendix; foreword and appendix by G. de Santillana. New York: McGraw-Hill, 1965. (P) (T)

Marcus, Rebecca. Galileo and Experimental Science. New York: Watts, 1961.

Santillana, Giorgio de. The Crime of Galileo. Chicago: University of Chicago Press, 1955 (P); French trans. by F.M. Rosset, Galilée. Paris: Laffont, 1968.

Seeger, Raymond J. Men of Physics: Galileo Galilei, His Life and Works. Oxford and New York: Pergamon Press, 1966. (P) (19-22, T)

Includes extracts from Galileo's writings.

Josiah Willard GIBBS (1839-1903)

Leesburger, Benedict A., Jr. Josiah W. Gibbs, American Theoretical Physicist. New York: Watts, 1963. (12-16)

Rukeyser, Muriel. Willard Gibbs. Garden City, N.Y.: Doubleday, 1942; New York: E.P. Dutton, 1964. (P) (20-24)

Wheeler, Lynde Phelps. Josiah Willard Gibbs, The History of a Great Mind. New Haven and London: Yale University Press, 1951; Hamden, Conn.: Shoestring Press. (20-24)

Werner HEISENBERG (1901-)

Heisenberg, Werner. Physics and Beyond, Encounters and Conversations (trans. from German). New York: Harper & Row, 1971.

Hörz, Herbert. Werner Heisenberg und die Philosophie. Berlin: VEB Deutscher Verlag der Wissenschaften, 1968.

Hermann von HELMHOLTZ (1821-1894)

Königsberger, Leo. Hermann von Helmholtz. Braunschweig: Vieweg, 1902. English trans., abridged, by F.A. Welby. Oxford: Clarendon Press, 1906; New York: Dover Publications, 1965. (P) (T)

Ebert, Hermann. Hermann von Helmholtz. Stuttgart: Wissenschaftliche Verlagsgesellschaft, M.B.H., 1949.

Christian HUYGENS (1629-1695)

Bell, A.E. Christian Huyghens and the Development of Science in the Seventeenth Century. London: Arnold, 1947. (20-24)

Dijksterhuis, E.J. Christiaan Huygens. Haarlem: Bohn, 1951.

James Prescott JOULE (1818-1889)

Wood, Alex. Joule and the Study of Energy. London: Bell, 1925.

William Thomson, Lord KELVIN (1824-1907)

Gray, Andrew. Lord Kelvin, An Account of His Scientific Life and Work. London: Dent & New York: Dutton, 1908.

Russell, Alexander. Lord Kelvin. London: Blackie, 1938.

James Clerk MAXWELL (1831-1879)

Campbell, Lewis and Garnett, William. The Life of James Clerk Maxwell. 2nd ed.; London: Macmillan, 1882; New York: Johnson Reprint Corp., 1969 (with a new introduction by Robert Kargon). (L)

May, Charles P. James Clerk Maxwell and Electromagnetism. New York: Watts, 1962. (12-16)

Julius Robert MAYER (1814-1878)

Hell, B. J. Robert Mayer und das Gesetz der Erhaltung der energie. Stuttgart; 1925.

Schmolz, Helmut & Weckbach, Hubert. Robert Mayer, sein Leben und Werk in Dokumenten. Weissenhorn: A.H. Konrad, 1964.

Schütz, Wilhelm. Robert Mayer. Leipzig: Teubner, 1969.

Robert A. MILLIKAN (1868-1953)

The Autobiography of Robert A. Millikan. Englewood Cliffs, N.J.: Prentice-Hall, 1950. (18-22)

Isaac NEWTON (1642-1727)

Andrade, E.N. da C. Sir Isaac Newton, His Life and Work. Garden City, N.Y.: Doubleday & Co., 1958. (P) (17-20)

Manuel, Frank E. A Portrait of Isaac Newton. Cambridge, Mass.: Harvard University Press, 1968. (T)

More, Louis Trenchard. Isaac Newton, A Biography. New York: Scribner, 1934; Dover Publications, 1962. (P) (T)

Palter, Robert, ed. The annus mirabilis of Sir Isaac Newton. Cambridge, Mass.: The M.I.T. Press, 1970. 328pp. (L)

A collection of papers on Newton's life and work presented at a conference.

Sootin, Harry. Isaac Newton. New York: Messner, 1955. (12-16)

Hans Christian OERSTED (1777-1851)

Dibner, Bern. Oersted and the Discovery of Electromagnetism. Norwalk, Conn.: Burndy Library, 1961. (L)

A brief account with excellent illustrations and bibliography.

Max Karl Ernst Ludwig PLANCK (1858-1947)

Hartmann, Hans. Max Planck als Mensch und Denker. Thun: Ott Verlag, 1953.

Kretzschmar, Hermann. *Max Planck als Philosoph*. Munich: E. Reinhardt, 1967.

Planck, Max. *Wissenschaftliche Selbstbiographie*. Leipzig: Barth, 1948. English trans. by F. Gaynor, *Scientific Autobiography*. New York: Philosophical Library, 1949. (20-24)

Wilhelm Conrad RÖNTGEN (1845-1923)

Glasser, O. *Wilhelm Conrad Röntgen und die Geschichte der Röntgenstrahlen*. Berlin/Göttingen/Heidelberg: Springer, 2 Aufl. 1959.

Nitske, W.R. *The Life of Wilhelm Conrad Röntgen, Discoverer of the X-Ray*. Tucson: University of Arizona Press, 1971.

Benjamin Thompson, Count RUMFORD (1753-1814)

Brown, Sanborn C. *Count Rumford, Physicist Extraordinary*. Garden City, N.Y.: Doubleday & Co., 1962. (P) *Men of Physics: Benjamin Thomson, Count Rumford*. New York: Pergamon Press, 1967. (17-20)

Sparrow, W.J. *Count Rumford of Woburn, Mass*. New York: Thomas Y. Crowell Co., 1964; New York: Fernhill House. Published in England as *Knight of the White Eagle*. London: Hutchinson, 1964. (20-24)

Ernest RUTHERFORD (1871-1937)

Eve, A.S. *Rutherford*. London: Cambridge University Press, 1939. (L)

Andrade, E.N. da C. *Rutherford and the Nature of the Atom*. Garden City, N.Y.: Doubleday & Co., 1964. (P) (18-22)

McKown, Robin. *Giant of the Atom, Ernest Rutherford*. New York: Messner, 1962. (12-16)

Erwin SCHRÖDINGER (1887-1961)

Scott, William T. *Ernest Schrödinger, An Introduction to His Writings*. Amherst, Mass.: University of Massachusetts Press, 1967. (22-24)

Simon STEVIN (1548-1620)

Dijksterhuis, E.J. *Simon Stevin, Science in the Netherlands around 1600*. Trans. from the Dutch edition of 1943 by E.J. Dijksterhuis, R. Hooykaas, and M.G.J. Minnaert. The Hague: M. Nijhoff, 1970.

Historisch exakte, auf Grund vorzüglicher Quellenkenntnis ausgeführte vortreffliche Darstellung des Lebens und der Werke Simon Stevins. (H. Kangro)

Joseph John THOMSON (1856-1940)

Thomson, George Paget. *J.J. Thomson, Discoverer of the Electron*. Garden City, N.Y.: Doubleday, 1966. (P) Published initially as *J.J. Thomson and the Cavendish Laboratory in His Day*. London: Nelson, 1964. (19-22)

Thomson, J.J. *Recollections and Reflections*. London: G. Bell & Sons, 1936.

Allessandro VOLTA (1745-1827)

Dibner, Bern. *Alessandro Volta and the Electric Battery*. New York: Watts, 1965. (12-16)

Thomas YOUNG (1773-1829)

Wood, Alexander. *Thomas Young, Natural Philosopher, 1773-1829*. Cambridge: University Press, 1954. (T)

17. SOURCEBOOKS AND COLLECTIONS OF CASE STUDIES: GENERAL SCIENCE INCLUDING PHYSICS

Boorse, Henry A., and Motz, Lloyd. The World of the Atom. 2 vols. New York & London: Basic Books, 1966. (T)

Selections from about 100 authors, including not only theoretical and experimental atomic physics but also related selections from electromagnetism and chemistry. Biographical sketches are included.

Committee on the Study of History, Amherst College; Van R. Halsey, Jr. and Richard H. Brown, General Editors. Menlo Park, Calif: Addison-Wesley, 1971. (17-20)

A series of booklets containing source material interspersed with interpretive narrative. Includes two booklets by Jonathan Harris: Hiroshima: A Study in Science, Politics and the Ethics of War and Science and the American Character.

Conant, James Bryant, ed. Harvard Case Histories in Experimental Science. 2 vols. Cambridge, Mass.: Harvard University Press, 1948. Also available as 8 separate pamphlets. (18-22)

Includes: Robert Boyle's Experiments in Pneumatics (ed. Conant); Early Development of Concepts of Temperature and Heat (ed. D. Roller); Atomic-Molecular Theory (ed. L. Nash); Concept of Electric Charge (ed. D. Roller and D.H.D. Roller).

Duhem, P. Le Système du Monde, Histoire des Doctrines Cosmologiques, de Platon à Copernic. 10 vols. Paris; 1914- (L)

A monumental work and an unrivalled source book on cosmology, mechanics and physics in the Middle Ages. (G. Boutry)

Hall, Marie Boas. Nature and Nature's Laws, Documents of the Scientific Revolution. New York: Harper and Row, 1970. 381pp. (P) (20-24)

While it provides excerpts from a wide range of scientific writings in the 17th century, there is a large enough proportion of material on physics, chemistry, astronomy, and scientific societies to make this selection of interest to those readers concerned primarily with the history of physics.

Hurd, D.L., and Kipling, J.J. *The Origins and Growth of Physical Science*. 2 vols. 334 & 416pp. Baltimore, Md.: Penguin Books, 1964. (P) (20-24)

Based on *Moments of Discovery*, edited by G. Schwartz and P.W. Bishop (1958). Selections from 55 authors, about 40% chemistry, 40% physics, and 20% astronomy.

Klopfer, Leo E. *History of Science Cases*. Chicago: Science Research Associates, 1964-66. (O/P in U.S.A.) (15-18)

A series of 30 to 40-page booklets on topics such as "Frogs and Batteries," "Fraunhofer Lines," and "Air Pressure." There is also a Teacher Guide for each topic. The student edition contains questions and experiments to accompany the extracts from original sources.

Knight, David M. *Classical Scientific Papers, Chemistry*. London: Mills and Boon, 1968; New York: American Elsevier, 1968. 2nd Series, 1971. (L)

Reprints of papers on atomic theory (Dalton, Berzelius, Davy, Faraday), and on kinetic theory of gases (Herapath, Maxwell, Kelvin, Perrin) with brief editorial commentary. Despite the title this is a valuable collection for the history of *physics*.

Moulton, F.R., and Schrifferes, J.J. *The Autobiography of Science*. London: Murray; New York: Doubleday, 1960. (T, L)

More than a hundred extracts from the masterworks of all science, from the time of Hippocrates to the 1950s, each extract being preceded by a brief biography of the scientist concerned. A very good selection. (B. Gee)

Newman, James R. *The World of Mathematics*. 4 vols. New York: Simon & Schuster, 1956. (P)

The second volume includes selections from the writings of Galileo, Daniel Bernoulli, Moseley, Bragg, Schrödinger, Eddington, Heisenberg and other examples of mathematics applied to science.

Olson, Richard. *Science as Metaphor, The Historical Role of Scientific Theories in Forming Western Culture*. Belmont, Calif.: Wadsworth Pub. Co., 1971. (20-24, T)

Science Jackdaws. London: Jackdaw publications, 30 Bedford Square. (16-19)

Each Jackdaw is a package of separate sheets, facsimile reproductions of documents and illustrations, and short essays. Topics include "Newton and Gravitation," "The Discovery of the Galaxies," "Faraday and Electricity." There is also a series of History Jackdaws which includes "James Watt and Steam Power."

Shapley, Harlow; Wright, Helen; and Rapport, Samuel. Readings in the Physical Sciences. New York: Appleton-Century Crofts, Inc., 1948. 501pp. (P) (18-22)

Selections from about 60 authors on astronomy, geology, mathematics, physics, chemistry, and scientific method, mostly of a general nature; subject bibliographies at end of each section.

Tierney, Brian; Kagan, Donald; and Williams, L. Pearce, eds. Random House, Historical Issues Series. (18-22)

Booklets of 50 to 75 pages containing extracts of both original and secondary sources, with questions for discussion.

2 "Ancient Science - Metaphysical or Observational?"

#14 "The Scientific Revolution - Factual or Metaphysical?"

Young, Louise B. Exploring the Universe. 2nd ed.; Oxford and New York: Oxford University Press, 1971.

An anthology of ancient and modern writings on science, astronomy, and physics.

________. The Mystery of Matter. Oxford and New York: University Press, 1965. (P) (L)

An anthology of writings from 72 authors covering the development of physical concepts leading to the discovery of atomic energy and the structure of living matter.

18. SOURCEBOOKS AND COLLECTIONS OF CASE STUDIES: PHYSICS ONLY. [See also the books on special topics, Nos. 6-12 above.]

Beaton, K.B., and Bolton, H.A. A German Source-Book in Physics. Oxford: Clarendon Press, 1969. 315pp. (P) (22-24)

> 46 papers reprinted from 20th-century German physics journals (Planck, Einstein, Born, Schrödinger, Heisenberg, etc.) with footnotes explaining how to translate a few of the more difficult German words and phrases. The book is primarily intended for graduate students in physics who want to improve their German reading ability, but at the same time is an excellent sourcebook for the history of modern physics.

Clagett, Marshall. The Science of Mechanics in the Middle Ages. Madison: The University of Wisconsin Press, 1959. 711pp. O/P (L)

> Includes texts (in Latin), English translations, and commentaries on works of Jordanus de Nemore, Gerard of Brussels, the Merton and Paris groups, and others.

Drake, Stillman, ed. & trans. Discoveries and Opinions of Galileo. Garden City, N.Y.: Doubleday Anchor, 1957. (P) (17-22)

Haar, D. ter. Selected Readings in Physics. (See below, section 19) (20-24)

Harvard Project Physics. Readers 1 to 6. New York: Holt, Rinehart & Winston, 1970. (17-19)

> Good background reading for any student taking a science course. The readings selected are a collection of some of the best articles and book passages on physics. Some passages deal with historic events, some with what physicists do, and others deal with the philosophy of science. (B. Gee)

Koslow, Arnold. The Changeless Order, the Physics of Space, Time, and Motion. New York: Braziller, 1967. (19-22)

> A source book of important writings from antiquity to the present.

Magie, W.F. A Source Book in Physics. New York: McGraw-Hill, 1935. 620pp. Cambridge, Mass.: Harvard University Press. (L)

About 115 extracts, each 4 or 5 pages long with a 1-page biography; primarily 17th and 19th centuries.

Marion, Jerry B. A Universe of Physics. New York: Wiley, 1970.

Mirow, Bernd. Quellenheft zur Physik (Ergänzungshefte für den Physikunterricht, hrsg. K. Hahn). Braunschweig: G. Westermann, 1958.

Contains extracts of Galilei's Discorsi, of the writings of Oersted, J.R. Mayer, H. Hertz and W.C. Röntgen.

Nobel Lectures in Physics, 1901-62. 3 vols. New York: American Elsevier Pub. Co. (L)

Sambursky, S. Anthology of Physical Thought. London: Hutchinson, 1971.

Thayer, H.S., ed. Newton's Philosophy of Nature, Selections from His Writings. New York: Hafner, 1953. (P)

Toulmin, Stephen, ed. Physical Reality. New York: Harper, 1970. (P) (20-24)

Debates on the foundations of physics, from Mach vs. Planck to Bohr vs. Einstein and Bohm.

Wolff, Peter. Breakthroughs in Physics. London: New English Library, 1965; New York: The New American Library, 1965; Signet paperback. (20-24)

Various points in time have been considered as being major scientific breakthroughs. Wolff has chosen the works of six great scientists to illustrate this point: Archimedes on machines and buoyancy; Galileo on astronomy and free fall; Pascal on the vacuum; Newton on the mathematical principles of physics; von Helmholtz on the conservation of energy; and finally Einstein on relativity. (B. Gee)

Wright, Stephen. Classical Scientific Papers, Physics. London: Mills and Boon, 1964; New York: American Elsevier, 1965. (L)

Reprints of papers on atomic physics by J.J. Thomson, Rutherford, Chadwick, Moseley, Aston, C.T.R. Wilson, Cockcroft and Walton, Compton and Doan, and others.

Green, George and John T. Lloyd. Kelvin's Instruments and the Kelvin Museum. Glasgow: University of Glasgow, 1970.

Patterson, Elizabeth C. John Dalton and the Atomic Theory. Garden City, N.Y.: Doubleday, 1970; also Anchor Books, 1970 (P).

19. SERIES OF REPRINTS OF SCIENTIFIC WORKS

Ames, J.S. Scientific Memoirs. New York: American Book Co., 1898-1901. 15 vols.

Each volume contains extracts from papers of two or three scientists on a single topic in physics or chemistry, with short biographical notes.

Buturigaku Koten Ronbun Sosyo [Classical Papers in Physics]. A series of twelve volumes of sourcebooks in modern physics, prepared by the Group for History of Physics, published by Tokai University Press, 1969-1971.

The topics are: Heat Radiation and Quanta; The Light Quantum; Old Quantum Theory; Theory of Relativity; Kinetic Theory of Gases; Statistical Mechanics; Radioactivity; Electron; Models of Atom; Theory of Atomic Structure; Electron Theory of Metal; Magnetism. Each volume includes classical papers in Japanese translation, with a brief historical account by the editor of the volume. A description of the series has been published in English by Tetu Hirosige, in Japanese Studies in the History of Science, No. 8 (1969), pp. 17-20.

Cass Library of Science Classics. L.L. Laudan, General Editor, Social Science Building, University of Pittsburgh, Pittsburgh, Penn. (15213). (L)

Includes works of Huygens, Boyle, Hooke, Cavendish.

Classics of Science. Gerald Holton, General Editor. New York: Dover Publications, Inc. (T)

Collections of papers on topics such as radioactivity and transmutation, isotopes, high-energy accelerators, concepts of time, quantum mechanics, cosmology.

Collection History of Science. Éditions Culture et Civilisation, 115, avenue Gabriel Lebon, Brussels 16, Belgium. (L)

Includes works of d'Alembert, Ampere, the Bernoullis, Tycho Brahe, Copernicus, Euler, Faraday, Galileo, Gauss, Gilbert, Hauy, Helmholtz, Hooke, Huygens, Kepler, Laplace, Lavoisier, Newton, Ohm, etc.

History of Science Series. R.M. Young, Principal Adviser. Farnborough, Hants, England: Gregg International Publishers Ltd. (L)

The aim of this programme is to make available a number of works which reflect the impact of science on Victorian thought, with particular emphasis on evolution and the debate about man's place in nature. The list therefore includes classical works in psychology, theology, geology and evolutionary theory, along with collections of essays and biographical works . . . the list also includes examples of popular essays and debates about the role of scientific societies such as the British Association and the Royal Society in advancing knowledge.

Ostwalds Klassiker der exakten Wissenschaften. Split into two series after World War II: (1) Leipzig: Akademische Verlagsgesellschaft Geest & Portig; (2) Frankfurt am Main: Akademische Verlagsgesellschaft, later Braunschweig: Vieweg. (T, L)

Founded by the German chemist Wilhelm Ostwald (1853-1932), this is a famous series of booklets, each containing a German edition of a classic scientific book or collection of articles on a single topic, annotated in many cases by a well-known scientist working in the same field. The series has been continued after Ostwald's death by other editors. It includes works of d'Alembert, Ampere, Archimedes, Daniel Bernoulli, Jakob Bernoulli, Johann Bernoulli, Boyle, Brewster, Carnot, Celsius, Clairaut, Clapeyron, Clausius, Coulomb, Dulong, Einstein, Euler, Fahrenheit, Faraday, Fermat, Fourier, Fraunhofer, Fresnel, Gadolin, Galilei, Galvani, Gauss, Gay-Lussac, Green, Guericke, Helmholtz, Huygens, Jacobi, Kirchhoff, Lagrange, Laplace, v. Laue, Lebedew, Leibniz, Lomonosov, Magnus, Maxwell, Robert Mayer, Newton, Oersted, Ohm, Petit, Planck, Poiseuille, Rydberg, Smoluchowski, Sommerfeld, William Thomson (Kelvin), Volta, Weber, Wien.

The Royal Institution Library of Science. Sir William Lawrence Bragg, General Editor. New York: American Elsevier Pub. Co. (L)

A series of volumes (some forty in all) containing exact reproductions of Discourses in the Sciences delivered at the Royal Institution of Great Britain between 1851 and 1939. A ten-volume set of discourses on physics and chemistry has already appeared.

Selected Readings in Physics. D. ter Haar, General Editor. Elmsford, N.Y.: Pergamon Press, Inc. [10523] (20-24)

Each volume contains an introductory historical and/or expository survey of the subject, followed by reprints of the major papers. The selections are chosen for their suitability for use in university physics courses. Topics include: kinetic theory, the old quantum theory, wave mechanics, atomic spectra, lasers, X-ray and neutron diffraction, nuclear forces, applied group theory, transistors, early solar physics, theory of beta decay, electromagnetism.

A subseries, "Men of Physics," includes volumes reprinting and discussing the works of Eddington, Galileo, Landau, Langmuir, Lark-Horovitz, Rayleigh, and Rumford. (P)

Selected Reprints. A project of the AAPT Committee on Resource Letters. New York: American Institute of Physics, 335 East 45 Street. [10017] (T)

Each booklet includes 20th-century papers on a single topic, e.g., special relativity theory, nuclear structure, superconductivity, origin of the elements, liquid helium, cosmic rays, plasma physics. (P) See also Selected Papers on Cosmic Ray Origin Theories (ed. Stephen Rosen), published by Dover, a similar collection on a somewhat larger scale.

Source Books in the History of the Sciences. Edward H. Madden, General Editor. Cambridge, Mass.: Harvard University Press.

Includes reprints of earlier source books compiled by S.F. Magie (Physics), K.F. Mather & S.L. Mason (Geology), H.M. Leicester & H.S. Klickstein (Chemistry), H. Shapley (Astronomy) and new collections by D.J. Struik (Mathematics) and J. van Heijenoort (Logic).

<u>The Sources of Science</u>. Harry Woolf, General Editor. New York: Johnson Reprint Corporation. (L)

Includes works of Robert Boyle, Isaac Newton, Johannes Kepler, Michael Faraday, John Herschel, Thomas Young, James Clerk Maxwell, G.G. Stokes, Joseph Priestley, and many others, with new introductions by historians of science.

20. MICROFORM REPRODUCTIONS OF SCIENTIFIC WORKS

<u>History of Science 16th to 19th Century</u>. The editors of the journal <u>Manuscripta</u>. St. Louis, Missouri: The Pius XII Memorial Library. (L)

This is the third of a series of microfilm collections of rare and out-of-print scientific works preserved at the Chantilly Library. The set of 40 rolls contains 118 works, mostly by rather obscure scientists but including Boscovich, Boyle, and Haüy.

<u>Landmarks of Science</u>. Eds. Sir Harold Hartley, Duane H. D. Roller, I.B. Cohen, C.C. Gillispie, Edward Rosen. New York: Readex Microprint Corp. Micro-opaque cards. (L)

Will eventually include "more than 3,000,000 pages of source materials." Works of Ampere, Archimedes, the Bernoullis, Boyle, Cavendish, Copernicus, Davy, Descartes, Einstein, Euler, Faraday, Fourier, Fresnel, Gibbs, Gilbert, Guericke, Helmholtz, Henry, Huygens, Kelvin, many others.

21. OTHER SERIES OF BOOKS ON HISTORY OF SCIENCE

Abhandlungen und Berichte des Deutschen Museums. Berlin, München: R. Oldenburg; Düsseldorf: VDI Verlag, 1929-. (17-24)

Die Reihe enthält unter anderen technikgeschichtliche und physikgeschichtliche Abhandlungen, z.B. H. Netz, Vom Heronsball nach Calder Hall, 33 Jahrgang, Heft 1, 1965. (H. Kangro)

Boethius Texte und Abhandlungen zur Geschichte der exakten Wissenschaften, hrsg. Joseph Ehrenfried Hofmann, Friedrich Klemm, und Bernhard Sticker. Wiesbaden: Franz Steiner Verlag GMBH, 1963-. (T, L)

The series includes several monographs on topics in the history of physics.

Books on History of Science and Technology. Cambridge, Mass.: MIT Press. (T, L)

These are mostly books on the history of technology, but works of Boscovich and Priestley are included. Several are available in both paperback and hardcover editions.

Histoire de la Pensée. Paris: Hermann. (T)

Includes some distinguished works on the history of science by Alexandre Koyré and others.

History of Science Library. Michael A. Hoskin, Editor. New York: American Elsevier Pub. Co. (T, L)

A series of monographs and historical studies; 20 to 30 volumes are planned as a joint publication of Macdonald-Oldbourne of England and American Elsevier. Among volumes already published are books on atomism and conservation and on theories of light in the 17th century.

Grosse Naturforscher. Stuttgart: Wissenschaftliche Verlagsgesellschaft M.B.H.

The series, begun in 1948, includes biographies of Helmholtz, Galileo, Weber, and John Herschel.

Miscellaneous books. New York: Dover Publications, Inc. (T, L)

These are widely distributed, mostly paperback editions, and include works of Archimedes, Birkhoff, Boole, Borel, Born, Bridgman, Carnot, Copernicus, Descartes, Einstein, Euclid, Fourier, Galileo, Gauss, Gibbs, Gilbert, Helmholtz, Hertz, Hooke, Huygens, Lorentz, Mach, Maxwell, Nernst, Planck, Poincare, Rayleigh, J.J. Thomson, W. Thomson, etc. In addition there are a few hardcover editions, including works of the Bernoullis, Faraday, Maxwell, and Rayleigh. Several older books on the history of science have been reprinted. There are also collections of recent papers on topics such as mathematical analysis, quantum electrodynamics, and stochastic processes (see also the Classics of Science series listed above).

Publications in Medieval Science. Madison, Wisc.: The University Press. (L)

Includes texts, translations, and commentaries on works in mechanics.

The Rise of Modern Science Series. A. Rupert Hall, General Editor. New York: Harper & Row. Some available in the Harper Torchbook (paperback) series. (20-24)

Includes From Galileo to Newton 1630-1720 by A.R. Hall; volumes are planned on Physical Sciences in the 19th and 20th centuries.

22. JOURNALS CONTAINING ARTICLES ON HISTORY OF PHYSICS

[This list is rather short since it includes only those journals that specialize in history of physics. Most articles on the history of physics will be found, however, in other journals which cover the history of science in general—e.g., *Isis*, *The British Journal for the History of Science*, *Centaurus*, *Japanese Studies in History of Science*, etc.—or in journals devoted to physics education—e.g., the *American Journal of Physics*, *The Physics Teacher*, *Contemporary Physics*, *Physics Education*, etc. A comprehensive list of these other journals may be found at the beginning of the annual Critical Bibliographies published in *Isis*.]

Archive for History of Exact Sciences. C. Truesdell, Editor. Berlin/Heidelberg/New York: Springer-Verlag. (L)

Buturigakusi Kenkyu [*Studies in the History of Physics*]. Published since 1958 by the Japanese Group for History of Physics.

Historical Studies in the Physical Sciences. Russell McCormmach, Editor. Philadelphia: The University of Pennsylvania Press, Vol. I (1969). (T, L)

The Natural Philosopher. D. Gershenson and D. Greenberg, Editors. Blaisdell Pub. Co. 3 vols., 1963-64 (ceased publication).

23. PHYSICS TEXTBOOKS WRITTEN FROM AN HISTORICAL VIEWPOINT: GENERAL PHYSICS. (a) ENGLISH

Arons, Arnold. Development of Concepts of Physics. Reading, Mass.: Addison-Wesley, 1965. (18-20)

Cooper, Leon N. An Introduction to the Meaning and Structure of Physics. New York: Harper & Row, 1968. Abridged ed., 1970. (18-20)

Crawford, Franzo H. Introduction to the Science of Physics. New York: Harcourt, Brace & World, 1968. (18-20)

Easley, J.A., Jr., and Tatsuoka, Maurice M. Scientific Thought, Cases from Classical Physics. Boston: Allyn & Bacon, Inc., 1968. (P) (17-19)

See favorable review in Am. J. Phys., Vol. 37, p. 340.

Harvard Project Physics. The Project Physics Text. 6 vols. New York: Holt, Rinehart & Winston, 1970. (17-19)

Holton, Gerald. Introduction to Concepts and Theories in Physical Science. Reading, Mass. and London: Addison-Wesley Pub. Co., 1952. 650pp. 2nd ed. in preparation by S.G. Brush, to be published 1973. (19-21)

Holton, Gerald, and Roller, Duane H.D. Ed., Duane Roller. Foundations of Modern Physical Science. Reading, Mass.: Addison-Wesley Pub. Co., 1958. 782pp. (18-20)

Inglis, Stuart J. Physics, An Ebb and Flow of Ideas. New York: Wiley, 1970. (18-20)

A textbook which presents the historical development of certain fundamental concepts of physics in depth. (M.G. Ebison)

Karplus, R. Introductory Physics, A Model Approach. New York: Benjamin, 1969. (15-18)

Although not following a strictly historical development of physics, the author provides an interesting historical context for some aspects of the subject to illustrate the transient nature of most scientific theories and the extensive modifications that scientific models have undergone. (M.G. Ebison)

Kemble, E.C. Physical Science, Its Structure and Development. Vol. 1: From Geometric Astronomy to the Mechanical Theory of Heat. Cambridge, Mass.: The M.I.T. Press, 1966. Vol. 2: Atoms, Waves and Particles. Cambridge, Mass.: The M.I.T. Press, 1970 (Preliminary edition). (18-21)

These books serve as an introduction to physical science and attempt to bring out clearly the dual character of physical science as an expanding body of verifiable knowledge and as an organized human activity whose goals and values are major factors in contemporary society.

The first volume deals with the development of science from the astronomy of ancient Greece via the Copernican Revolution, Newtonian mechanics, the concept of energy and the steam engine, to the classical kinetic-molecular theory of heat. The second volume traces the development of quantum mechanical conceptions concerning the nature of matter and radiation from their historical roots in 19th-century chemistry, electromagnetic theory and optics. (M.G. Ebison)

March, R.H. Physics for Poets. New York: McGraw-Hill, 1970. (18-20)

Ripley, J., and Whitten, R. The Elements and Structure of the Physical Sciences. 2nd ed.; New York: John Wiley, 1969. (18-20)

Taylor, Lloyd W. Physics, the Pioneer Science. Boston: Houghton Mifflin, 1941; New York: Dover Publications, 1959. (P) (18-19)

(b) GERMAN

Kuhn, W. Physik. Braunschweig: Georg Westermann Verlag, Bd. I, 1967, 2. Aufl. 1970; Bd. II, 1968.

24. HISTORICAL TEXTBOOKS ON TOPICS IN PHYSICS

d'Abro, A. The Rise of the New Physics, Its Mathematical and Physical Theories (formerly entitled Decline of Mechanism). Vol. II, Quantum Theory. 2nd ed.; New York: D. Van Nostrand, 1951; New York: Dover Publications. (P) (21-24)

Angrist, Stanley W., & Hepler, L.G. Order and Chaos. New York: Basic Books, 1967. (20-23)

Mainly thermodynamics, some statistical concepts.

Cropper, William. The Quantum Physicists and An Introduction to Their Physics. New York: Oxford University Press, 1970. (P) (21-24)

This book presents not only the development of the formal basis of the theory but also the human endeavors of the scientists involved: Planck, Einstein, Bohr, Heisenberg, de Broglie, Schrödinger, Born and Dirac. (M.G. Ebison)

Friedman, Francis L., and Sartori, Leo. The Classical Atom. Reading, Mass.: Addison-Wesley Pub. Co., Inc., 1965. (P) (20-22)

A short account of atomic theories and experiments up to 1912.

O'Rahilly, Alfred. Electromagnetic Theory (formerly titled Electromagnetics [1938]). 2 vols. New York: Dover Publications. (T, L)

Slater, J.C. Concepts and Development of Quantum Physics. New York: Dover Pubs., 1969 (originally published under the title Modern Physics [1955]). (20-24)

The author presents historically the development of the ideas which contribute to the current understanding of atomic and molecular physics, and particularly quantum mechanics. (M.G. Ebison)

Tomonaga, Sin-itiro. Quantum Mechanics. 1. Old Quantum Theory. Amsterdam: North-Holland, 1962. 2. New Quantum Theory. Amsterdam: North-Holland, 1966. (22-24)

First published in Japanese in 1948.

25. EDUCATIONAL RESEARCH PROJECTS IN SCIENCE

Klopfer, Leopold E., and Cooley, William W. The History of Science Cases for High Schools in the development of student understanding of science and scientists. Journal of Research in Science Teaching, Vol. 1 (1963), pp. 33-47. (L)

> Conclusions: "The findings of the HOSC Instruction Project clearly demonstrate that the HOSC Instruction method is definitely effective in increasing student understanding of science and scientists when used in biology, chemistry, and physics classes in high schools. . . . Moreover, when students study under the HOSC Instruction Method, they achieve these significant gains in understanding of science and scientists with little or no concomitant loss of achievement in the usual content of high school science courses."

Korth, Willard William. "The Use of the History of Science to Promote Student Understanding of the Social Aspects of Science." Ph.D. Dissertation, Stanford University, 1968. (Dissertation Abstracts, Vol. 29, 1447-A, 1968; copies may be ordered from University Microfilms, Ann Arbor, Michigan; Order No. 68-15,069.) (L)

> This study represents an effort to develop a test suitable for assessing students' conceptions of the social aspects of science and to use this instrument to determine the change produced through an instructional unit based on a phase of the history of science. . . Six classes of biology students studied a two-week unit based on a case history of the cell theory. . . . The results of this study indicate that the experimental treatment had some effect in promoting understanding of the social aspects of science. However, while statistically significant changes were produced, the overall effect was far from impressive . . . no significant change was observed on that part of the test dealing with the social responsibilities of scientists. . . . It appears that many students have built up a conception of science and the scientific enterprise over several years of science instruction at lower levels which is relatively resistant to change.

Lavach, John Francis. "An In-Service Program in the Historical Development of Selected Physical Science Concepts." Ed. D. Dissertation, Duke University, 1967. (Dissertation Abstracts, Vol. 28, 4513-A; copies may be ordered from University Microfilms, Ann Arbor, Michigan; Order No. 68-6899.) (L)

> The author has developed a "Test on the historical development of science" which he used to measure teachers' gains in understanding history of science in his program.

Merkin, Melvin. "The Effect of Case Studies in Science upon College General Education Courses in Physical Science." Ed. D. Dissertation, Boston University School of Education, 1967. (Dissertation Abstracts, Vol. 29 (1969), p. 4353-A; copies may be ordered from University Microfilms, Ann Arbor, Michigan; Order No. 69-7821.) (L)

> The experiment was designed to test three null hypotheses of no significant differerences between the mean scores of the experimental and control groups on the *Test on Understanding Science*, *Test of General Physical Science*, and *Perception of Science and Scientists Test* (PSST) ... *Conclusions*: (1) Students in classes in which case studies [e.g., Klopfer's "Fraunhofer Lines"] were used emerged with a better understanding of science and scientists than students who were not in classes using these materials. (2) Time spent on a case study did not adversely affect performance on a test of the normal physical science content of the course. (3) Students in classes in which case studies were used emerged with more positive and constructive attitudes toward science and scientists. . . .

Peterson, Donald Gordon. "An Experimental Comparison of Case Histories with Conventional Materials in Teaching a College General Education Course in Science." Ph.D. Dissertation, University of Minnesota, 1968. (Dissertation Abstracts, Vol. 29, 2155-A, 1969; copies may be ordered from University Microfilms, Ann Arbor, Michigan; Order No. 68-17,731.) (L)

> In the experimental treatment, the nature of science and scientific research and other content were presented through case histories while in the control treatment scientific facts and generalizations were presented through semi-popular current readings. The

course content in both treatments centered on atomic-molecular concepts. . . . Three tests were developed and used as pre- and post-tests: Test of Scientific Information (TSI), Test of Scientific Methods (TSM), and Test of Scientific Attitudes (TSA). [Conclusion: the experimental groups did better on the TSM and TSA but the control group did better on the TSI.]

26. BIBLIOGRAPHIES IN HISTORY OF SCIENCE

Bierens de Haan, D. Bibliographie Neerlandaise Historique Scientifique des Ouvrages Importants dont les Auteurs sont nes aux 16e, 17e et 18e Siècle, sur les Sciences Mathematiques et Physiques, avec leurs Applications. Rome: 1885; reprinted by de Graef, Nieuwkoop, 1965. (L)

Bulletin Signaletique. Centre national de la recherche scientifique, Sec. 22: "Histoire des Sciences et des techniques." Three issues per year. (L)

"Critical Bibliographies of the History of Science and Its Cultural Influences," now published annually in Isis as a separate number, under the supervision of Mr. John Neu, The Memorial Library, The University of Wisconsin, Madison, Wisconsin. (L)

> The first 92 bibliographies cover publications in the history of science from 1912 through 1966, arranged primarily by the historical period referred to, and by subject within each period. Approximately 2500 entries each year. Cumulative index in preparation.

Catalogue of Reprints of Scientific Papers in the Pauli Collection. CERN (European Organization for Nuclear Research). Geneva, Switzerland: 1969. 470pp. (L)

> Citations of more than 10,000 research papers in 20th-century physics, representing Pauli's selection of what was worth keeping out of the reprints sent to him.

Cohen, I. Bernard. "Some Recent Books on the History of Science," pp. 627-656 in Roots of Scientific Thought. Eds. P.P. Wiener and A. Noland. New York: Basic Books, 1957. (T)

Hall, Marie Boas. History of Science. 2nd ed. 1964. Publication Number 13 of the Service Center for Teachers of History, American Historical Association, 400 A Street SE, Washington, D.C. [20003]. (T)

> A survey of books available for teachers of high school and college courses in history of science.

Crombie, A.C., and Hoskin, M.A., eds. History of Science, An Annual Review of Literature, Research, and Teaching. Cambridge: Heffer & Sons, Vol. 1-, 1962-. (L)

Includes review articles on various topics with extensive bibliographies.

Klopfer, Leopold E. Paperbound Books in the History and Philosophy of Science. Revised and updated by Franklin G. Fisk, 1968. Mimeographed copies available from the Department of Secondary Education, The Pennsylvania State University, University Park, Penn. [16802]. (T)

Lipetz, Ben-Ami. A Guide to Case Studies of Scientific Activity. Carlisle, Mass.: Intermedia, Inc., 1965. (T, L)

Abstracts of about 400 articles and books containing case histories in science.

Neu, John. "The History of Science." Library Trends, Vol. 15, pp. 776-792 (1967). (L)

A survey of bibliographies indexing current publications in the history of science.

Poggendorff, J.C. Biographisch-Literarisches Handwörterbuch zur Geschichte der exacten Wissenschaften, enthaltend Nachweisungen über Lebensverhältnisse und Leistungen von Mathematikern, Astronomen, Physikern, Chemikern, Mineralogen, Geologen usw. aller Völker und Zeiten. Leipzig: J.A. Barth, 1863-date. New York: Johnson Reprint Corp., Vol. 5 (1904-1922); Vol. 6 (1923-1931). (L)

Although not as complete in its coverage of 19th-century periodical literature as the Royal Society Catalogue, this continuing series covers 18th and 20th century scientists, giving biographical data as well as books and articles they have published. (L)

"Resource Letters," published in the American Journal of Physics. Reprints available from the American Institute of Physics, 335 E. 45 St., New York, N.Y. [10017]. (T)

Almost all of these include some reference to classic papers; of special interest for the history of physics are the letters on "Evolution of the Electromagnetic Field Concept," "Evolution of Energy Concepts from Galileo to Helmholtz," "Science and Literature," "Philosophical Foundations of Classical Mechanics," and "Collateral Reading for Physics Courses."

Rider, K.J. History of Science and Technology. A Select Bibliography for Students. London: The Library Association. 1st ed., 1967; 2nd ed., 1970. (T)

Royal Society of London. Catalogue of Scientific Papers. Published in 19 volumes plus 4 volumes of (incomplete) subject index. London: 1867-1925. Reprinted by Scarecrow Press (Metuchen, N.J.) and Kraus Reprint Corp. (N.Y.). (L)

Lists almost all scientific papers (but not books) published from 1800-1900. Continued up to 1914 by the International Catalogue of Scientific Literature.

Russo, F. Elements de Bibliographie de l'Histoire des Sciences et des Techniques. Paris: Hermann, 1954; 2nd ed., 1969. 230pp.

Sarton, George. A Guide to the History of Science. Waltham, Mass.: Chronica Botanica Co., 1952. (T, L)

Das Buch enthält ausführliche Verzeichnisse über Bücher zur Geschichte der Naturwissenschaft, Physik, Technik, etc., und über einschlägige Zeitschriften. (H. Kangro)

27. DESCRIPTIONS OF MANUSCRIPT COLLECTIONS AND ARCHIVES

Albergotti, J.C. "West European Scientific Museums: A Survey." American Journal of Physics, Vol. 39 (1971), pp. 243-253. (T)

Denecke, Ludwig. Die Nachlässe in den Bibliotheken der Bundesrepublik Deutschland (Verzeichnis der schriftlichen Nachlässe in deutschen Archiven und Bibliotheken, Bd.2). Boppard am Rhein: Harald Boldt Verlag, 1969. (L)

Gee, Brian. "Musée d'Histoire des Sciences (Génève)," etc. Mimeographed working papers for the International Working Seminar, 1970. (T)

Hermann, Armin. "History of Science and Resource Centers in the German Speaking Countries." Mimeographed working paper for the International Working Seminar, 1970. (T)

Hoskin, Michael. "Scientific Archives, Museums, Libraries and Other Collections in the United Kingdom." Mimeographed working paper for the International Working Seminar, 1970. (T)

Kuhn, Thomas S.; Heilbron, John L.; Forman, Paul L.; and Allen, Lini. Sources for History of Quantum Physics, An Inventory and Report. Philadelphia: The American Philosophical Society, 1967 (Volume 68 of its Memoirs). 176pp. (L)

> Report on a project directed by T.S. Kuhn, sponsored by the American Physical Society and the American Philosophical Society; list of 175 tape-recorded and transcribed interviews with almost 100 physicists who participated in the development of quantum physics; also lists of letters, draft lectures, research notes and institutional records. The materials themselves are deposited at Philadelphia, Berkeley, and Copenhagen.

Verzeichnis der Autographensammlung von Professor Dr. Ludwig Darmstaedter. Berlin: J.A. Stargardt, 1909. (Schriften der Königlichen Bibliothek Berlin) (L)

> Dieser Katalog enthält nur die Ewerbungen bis 1909 aufgezeichnet. (H. Kangro)

Warnow, Joan Nelson. <u>National Catalog of Sources for History of Physics</u>. Report No. 1: A selection of Manuscript Collections at American Repositories, New York: Niels Bohr Library, Center for History and Philosophy of Physics, American Institute of Physics, 1969. (L)

Summaries of 103 collections, with name index and list of finding aids and other materials in the Niels Bohr Library.

28. FILMS ON INDIVIDUAL PHYSICISTS

A New Reality [Bohr] (51 minutes). Color. Statens Filmcentral, Denmark, for OECD, in association with the International Council for Educational Films. English version prepared by Educational Foundation for Visual Aids (EFVA); distributed by EFVA, 33 Queen Anne St., London W1M OAL, England. With teachers' notes. Also available from International Film Bureau, 323 South Michigan Avenue, Chicago, Illinois [60604].

> The film traces the discovery of the structure of the atom with particular emphasis on the work of Niels Bohr. The first part traces the evolution of the theories of atomic structure from the end of the 19th century, from the first unequivocal information concerning the electron, the clear evidence that the positive charge in the atom is concentrated in a small central nucleus and that the atom radiates light in bursts. The main part is devoted to the work of Niels Bohr and the final sequences provide an historical survey of earlier beliefs and touch briefly on the philosophical implications of Bohr's concept. Age: 16 years and over. [From EFVA catalog]

The Small World of Niels Bohr (60 minutes). State University of New York, Educational Communications Office, Room 2332, 60 East 42nd Street, New York, N.Y. [10017]. 1966. With Edward Teller.

> Account of the revolution in scientific thought effected by Bohr's wave-particle dualism concept; describes influence of his ideas on modern science. (W.R. Riley, Am. J. Phys. 36 #6)

The Large World of Albert Einstein (60 minutes). State University of New York, Educational Communications Office, Room 2332, 60 East 42nd Street, New York, N.Y. [10017] 1966. With Edward Teller.

> Basic geometric laws describing earth-bound time-distance relationships are extended to relative motion. Gauges impact of theory of relativity on 20th century scientific thought. (W.R. Riley, Am. J. Phys. 36 #6)

Prelude to Power (25-1/2 minutes). Greenpark Productions, for OEEC and Educational Foundation for Visual Aids

(EFVA); distributed by EFVA, 33 Queen Anne Street, London W1M OAL, England. With teachers' notes.

This film presents the story of Faraday and the Induction Ring and is intended for non-specialists in the middle forms of Secondary Schools. Dealing with the life and work of Michael Faraday, especially in the field of electro-magnetism, the film shows the experiments that led to the discovery of the principle of the induction ring and emphasizes the importance of this in the development of the dynamo, the source of electric power in the modern world. Age: 13-16 years. (From EFVA catalog)

The World of Enrico Fermi (46 minutes). Harvard Project Physics. Holt, Rinehart & Winston, P.O. Box 2334, Grand Central Station, New York, N.Y. [10017]

A documentary film, primarily for pedagogical purposes, produced by Harvard Project Physics with a grant from the Ford Foundation. Its main aim is to give the viewer an appreciation for the life and contribution of a nearly contemporary physicist, one who was widely honored and loved, and whose work helped to transform not only physics and the style of doing science, even the course of history itself. . . . The viewer meets a number of the foremost scientists in documentary film footage or stills. He sees some of the equipment Fermi used, and glimpses the way Fermi made teams work so well. He sees the locations at which Fermi was active (including footage of laboratories in Rome, Columbia, Chicago, and Los Alamost). But in following this work, e.g., from the discovery of slow-neutron-induced radioactivity to the nuclear reactor and the A-bomb, he also finds the kind of questions raised which are on the minds of students and the public today: those concerning the relations between physics, technology, and social concerns. (From teacher notes prepared by Project Physics)

Galileo (13-1/2 minutes). Marshall Clagett. Coronet Instructional Films, 65 East South Water Street, Chicago, Ill. [60601] (12-18)

Throughout his life, Galileo fought for the right of science to question tradition. Shot in Italy, at locations associated with Galileo, the film recreates some of the dramatic elements of that struggle. We see how he disproves earlier theories of Aristotle by demonstrating his laws of falling bodies, how he verified the astronomical theories of Copernicus and how, in spite of strong opposition, he continued to make new discoveries in the realm of physical science. Chris Award, Eighth Annual Columbus Film Festival. Junior High, Senior High. (From Coronet catalog)

Isaac Newton (13-1/2 minutes). Carl B. Boyer. Coronet Instructional Films, 65 East South Water Street, Chicago, Illinois [60601]

How the genius of Sir Isaac Newton significantly changed the course of several branches of the physical sciences and of mathematics is the theme of this film. Newton's researches into the binomial theorem, the differential and the integral calculus, his theory of light, and his law of gravitation and laws of motion are dramatically reenacted as they relate to his life and work in England. Junior High, Senior High. (From Coronet catalog)

Young and the Wave Theory of Light (21 minutes). Color. Sir Lawrence Bragg at the Royal Institution. Anvil Films in association with Educational Foundation for Visual Aids (EFVA); distributed by EFVA, 33 Queen Anne Street, London W1M OAL, England.

Sir Lawrence outlines the two contesting theories of light, the particles theory and the wave theory, which were held in the 18th century. He follows with a description and demonstration of Thomas Young's famous pinhole experiment; an experiment which validated the wave theory. Sir Lawrence turns to diffraction, showing its effects with the use of Young's original Wave Trough of 1800, and explaining diffraction fringes in a series of demonstrations. Age: 12 years and over. (From EFVA catalog)

29. FILMS ON TOPICS IN THE HISTORY OF PHYSICS

ATOMIC PHYSICS, X-RAYS, RADIOACTIVITY
(See also #28 under Bohr)

Atomic Physics (80 minutes). Universal Education and Visual Arts, 221 Park Avenue South, New York, N.Y. [10003]. J.A. Rank Organization, Ltd. In 5 parts: The Atomic Theory; Rays from Atoms; The Nuclear Structure of the Atom; Atom Smashing; Uranium Fission; Nuclear Energy. See details below under these titles.

W.R. Riley (Am. J. Phys. 36 #6) says this is a "superb resumé of work of atomic and nuclear physicists from Dalton's early 19th century approach to post-Bikini peaceful efforts."

Atom-Smashing (22 minutes). Universal Education and Visual Arts, 221 Park Avenue South, New York, N.Y. [10003] (17-20)

Sr. High, College. The work of the Curies and James Chadwick in discovery of the neutron is projected; splitting of the lithium atom by Cockcroft and Walton is discussed; Einstein explains his theory of mass and energy. The cyclotron and its use is illustrated. (From UEVA catalog)

The Atomic Theory (10 minutes). Universal Education and Visual Arts, 221 Park Avenue South, New York, N.Y. [10003]

Sr. High, College. A general approach starting with the basic theory proposed by Dalton in 1808, and outlining progress in atomic study during the 19th century, including Faraday's electrolysis experiments and Mendeleev's Periodic Table. (From UEVA catalog)

Conquest of the Atom (22 minutes). Color. Realist, for Mullard and Educational Foundation for Visual Aids (EFVA); distributed by EFVA, 33 Queen Anne Street, London W1M OAL, England. With teachers' notes.(14-18)

This film was produced with the assistance of the U.K. Atomic Energy Authority and tells the story of the structure and splitting of the atom. It begins in 1895 with the work of J.J. Thomson in the Cavendish Laboratory in Cambridge, which by 1897 was to disprove the conception of the atom as a minute, solid, indivisible particle. The next section of the film deals with the work of Rutherford, Thomson's pupil, which resulted in the first splitting of the nitrogen atom in 1919. The story continues with Sir James Chadwick's work, the discovery of the neutron and the harnessing of atomic power. (From EFVA catalog)

<u>The Day Tomorrow Began</u> (30 minutes). Color. Produced by U.S. Atomic Energy Commission, Argonne Nat. Lab. For sale by National Audiovisual Center (GSA), Washington, D.C. [20409]. Available on free loan from 11 AEC film libraries in the U.S.A. (See AEC film catalog for addresses.)

This historical film tells the story of the building and testing of CP-1 (Chicago Pile-1), the first atomic pile, and the work of the brilliant scientific team, led by Dr. Enrico Fermi, which ushered in the Atomic Age behind a cloak of wartime security under the stands of Stagg Field, Chicago, December 2, 1942. By interview, historical footage, paintings, etc., the film takes us on a step-by-step re-enactment of the famous event——beginning with the arrival of the first refugee scientists in 1939, to the dramatic hours in late 1942 when control rods were pulled out of CP-1 an inch at a time, to achieve the first sustained chain reaction. Interviews are conducted with some of the members of the team and people closely associated with them——John Wheeler, Mrs. Laura Fermi, Glenn Seaborg, Leslie Groves. (From AEC catalog)

<u>The Discovery of Radioactivity</u> (15 minutes). Color. Institut für Film und Bild, for OECD, in association with the International Council for Educational Films. English version prepared by Educational Foundation for Visual Aids (EFVA); distributed in England by EFVA, 33 Queen Anne Street, London W1M OAL, England. With teachers' notes. (13-18)

An historical survey of the discovery and progressive developments leading to our present knowledge of radioactivity. The work of Röntgen, Henri Becquerel, Marie Curie, Elster and Geitel and the analysis of the phenomena by Rutherford is portrayed in this film. The closing sequences deal with more recent discoveries. (From EFVA catalog)

The Nuclear Structure (19 minutes). Universal Education and Visual Arts, 221 Park Avenue South, New York, N.Y. [10003]

Sr. High, College. Reconstructs the early work of Becquerel and the Curies and radioactivity, and explains Rutherford's theory of the nuclear structure of the atom. Animated diagrams explain H.G. Moseley's work. (From UEVA catalog)

Rays from Atoms (12 minutes). Universal Education and Visual Arts, 221 Park Avenue South, New York, N.Y. [10003]

Sr. High, College. Demonstrates early work with cathode rays and discovery of the electron; how positive rays were discovered and their nature established; the work of Roentgen with X-rays; the work of Sir Joseph Thomson. (From UEVA catalog)

Strangeness Minus Three (45 minutes). BBC-TV. N. Samios, M. Gell-Mann, and Y. Ne'eman; distributed by Peter M. Robeck & Co., 230 Park Avenue, New York, N.Y. [10017]

The story behind the search for the elusive "omega-minus" particle. Philosophical implications explained by Feynman.

Uranium Fission (24 minutes). Universal Education and Visual Arts, 221 Park Avenue South, New York, N.Y. [10003]

Sr. High, College. A general approach leading to the discovery of uranium fission is reviewed and explained. Developments in the making of the atomic bomb are outlined. Peacetime research and a hopeful look at the future are included. (From UEVA catalog)

ELECTRICITY AND MAGNETISM

(See also under Faraday in #28)

The Development of Electrochemistry (19 minutes). Color. Statens Filmsentral, Norway, for OECD, in association with the International Council for Educational Films. English version prepared by Educational Foundation for Visual Aids (EFVA); distributed in England by EFVA, 33 Queen Anne Street, London W1M OAL, England. With teachers' notes. (15-19)

The film provides an historical survey of one of the fundamental discoveries in modern science, that of electric current, and covers the work of Volta, Sir Humphrey Davy, Oersted, Faraday, and Van't Hoff. (From EFVA catalog)

The Discovery and Generation of Electromagnetic Waves (Part 1 of Electromagnetic Waves) (20 minutes). Color. Realist for Mullard and E.F.V.A. distributed by Educational Foundation for Visual Aids, 33 Queen Anne Street, London W1M OAL, England. Includes teachers' notes. (17-20)

The history of the discovery of the electromagnetic spectrum is outlined and the film illustrates the generation of frequencies from radio- to gamma-waves by classical experiments and animated diagrams. Both wave and photon concepts are introduced. The film concludes with a cross-section of the modern uses of electro-magnetic radiation. (EFVA catalog) Rec. by M.G. Ebison.

Magnetism. In 3 parts (15-1/2 + 16-1/2 + 17-1/2 minutes). Color. Sir Lawrence Bragg at the Royal Institution. Anvil Films in association with Educational Foundation for Visual Aids (EFVA); distributed by EFVA, 33 Queen Anne Street, London W1M OAL, England. Teachers' notes.

1. Sir Lawrence begins with a brief survey of the history of magnetism and illustrates some of its familiar effects. A description of the fundamental properties of magnetic materials is followed by demonstrations with the lodestone, a modern magnet, and a magnetic compass. Magnetic attraction, repulsion, and induction are next described and illustrated. Finally, Sir Lawrence discusses Faraday's contribution, reconstructing original experiments and supplementing these with demonstrations illustrating the effects of the magnetic field. Age: 12 years and over.

2. The concept of the magnetic field is further developed by Sir Lawrence in his description of Oersted's great discovery of the early 19th century—the relationship between magnetism and electricity. The proof of this relationship is shown in a number of demonstrations. Sir Lawrence describes the magnetic properties of iron, and in a series of experiments illustrates its behaviour when subjected to the influence of a magnetic field. Age: 12 years and over.

3. Sir Lawrence first describes and contrasts the two classes of magnetic bodies——ferromagnetics and paramagnetics——and demonstrates the relationship between magnetic properties and temperature. He discusses the earth's magnetism and illustrates Wegener's novel theory of the history of the continents. Sir Lawrence shows how more recent knowledge of magnetism has fully validated Wegener's theory. Age: 12 years and over. (From EFVA catalog)

The Story of Electricity--The Greeks to Franklin (13-1/2 minutes). Duane H.D. Roller. Coronet Instructional Films, 65 East South Water Street, Chicago, Illinois [60601] (11-14)

Reenactments of the key advances in man's knowledge of electricity are told in the actual words of the discoverers, from the early Greek's elektron, or amber, to Benjamin Franklin's single-fluid theory. The ideas, methods and inventions of William Gilbert, Stephen Gray, Francis Hauksbee, Pieter van Musschenbroek and finally Franklin are illustrated. (From Coronet catalog)

The Story of Magnetism (13-1/2 minutes). Universal Education and Visual Arts, 221 Park Avenue South, New York, N.Y. [10003]

Intermediate, Jr. High. Dramatizes in authentic historical setting the story of magnetism from the discovery of loadstone, through the experiments of Gilbert, Oersted, and Faraday, to the latest theories of scientists. The film explains the kinds of magnets, sources of magnets, magnetic fields, and electromagnets. (From UEVA catalog)

GRAVITY

The Force of Gravity. McGraw-Hill.

Film demonstrates the nature of gravitation and discusses attempts of Newton and others to understand it. (Rec. by M.G. Ebison)

The Law of Gravitation, An Example of Physical Law (55 minutes). The first of Richard Feynman's Messenger Lectures at Cornell, "Character of Physical Law."

Education Development Center, Film Librarian, 39 Chapel Street, Newton, Mass. [02160]. 1964.

Brief story of its discovery, consequences, and effects on history of science; refinements and relation to other laws of physics. (W.R. Riley, Am. J. Phys. 36 #6)

OPTICS

(See also under Young in #28)

From Leeuwenhoek to Electronic Microscope (18 minutes). N. V. Multi-film, Steynlaan, 1, Hilversum, The Netherlands, 1953.

Measurement of the Speed of Light. Made for A.A.P.T. by McGraw-Hill, New York.

Fizeau's toothed wheel and Michelson's modification. (Rec. by M.G. Ebison)

Young and the Wave Theory of Light (see above, category #28).

THEORETICAL PHYSICS

The Evolution of Physical Ideas (49 minutes). The State University of New York, Educational Communications Office, Room 2332, 60 East 42nd Street, New York, N.Y. [10017]. 1966.

P.A.M. Dirac's personal approach to theoretical physics; suggests that physicists' attempts to improve existing theories involve a search for mathematical beauty. (W.R.Riley, Am. J. Phys. 36 #6.

30. OTHER AUDIOVISUAL AIDS

Anniversaries of Great Personalities and Events. Tape Recordings available on free loan from Radio & Visual Information Division, UNESCO, Place de Fontenoy, Paris-7e, France.

The following tapes (each about 30 minutes) are available:

$E = mc^2$ (Albert Einstein)	Tape No. 9,445
The Enigmatic Force (Isaac Newton)	9,942
The Gentle Man of Science (Michael Faraday)	10,578
Marie Curie	

(B. Gee)

Famous Scientists Wallcharts. Set of 21 charts. Paper, mounted on metal rims, 20" x 30", b/w. Central Electricity Generating Board, available from Educational Foundation for Visual Aids, 33 Queen Anne Street, London W1M OAL, England. (15-18)

These charts contain photographs, illustrations and text concerning the work of the following scientists: Archimedes, Boyle, Priestley, Coulomb, Lavoisier, Avogadro, Davy, Ohm, Faraday (4 charts), Joule, Crookes, Röntgen, Fleming, Becquerel and the Curies, Thomson, Rutherford (2 charts), and Aston.

Galileo and His Times (wall chart, 40 in. by 30 in.). Pictorial Charts, 181, Uxbridge Road, London W7, England. (Cite Ref. E26) (16-19)

Shows the world in which Galileo lived and the significance of his discoveries.

History of Magnetism. Color filmstrip (28 frames). Mullard Educational Service; order from The Slide Centre Ltd., Portman House, 17 Broderick Road, London M.S.W. 17, England. (11-18)

This filmstrip outlines the development of magnetic principles from ancient time to present day. Terrestrial magnets, electromagnetism, applications and properties of magnetic materials are all dealt with. (Mullard catalog)

Klassische Experimente der Physik von H.J. Bersch und K.H. Wiederkehr. Hamburg: Rohwolt, 1970.

Wiedergabe einer Reihe von Experimenten mit historischem Kommentar, die Frühjahr 1970 im

Fernsehen (3. Programm des Nordwestdeutschen Rundfunks) gesendet wurden. Die historischen Gedankengänge und der Hauptteil der physikalischen Experimente stammen von K.H. Wiederkehr. (H. Kangro)

Michael Faraday. Filmstrip in b/w (18 frames), with teachers' notes. The Electricity Council, E.D.A. Division, distributed by Educational Foundation for Visual Aids, 33 Queen Anne Street, London W1M OAL, England. (13-20)

This strip describes some of the experiments which led to Faraday's great electrical discoveries. A few of Faraday's own sketches are included. The notes give a brief summary of his life. (E.F.V.A. catalog)

Physics and Painting. A synchronized slide-lecture presentation. Available in U.S.A. on loan from the National Gallery of Art, Washington, D.C., free except for postage and insurance charges. 32 slides and 30-minute recording. (18-22)

The changes in thought about the structure of the physical world connected with names such as Copernicus, Newton, Einstein and other scientists are reflected in artistic changes from the Middle Ages to the present. How artists have painted volume, space, motion, time, weight, and light at different times in history, illustrated in works by major artists. (N.G.A. brochure)

Portraits of Famous Physicists. Portfolio of 12 portraits. 10" x 15", b/w. New York: Pictorial Mathematics, 1942. Reprint. Available from Educational Foundation for Visual Aids, 33 Queen Anne Street, London W1M OAL, England. (15-18)

The portfolio contains portraits of Newton, Galileo, Huygens, Ampère, Faraday, Fresnel, Rowland, Joule, Clausius, Hertz, Gibbs, and Maxwell, with biographical sketches by Professor H. Crew.

Posters on Great Physicists (11" x 14") by David E. Newton. Portland, Maine: J. Weston Walch, 1967.

Schallplatten Stimme der Wissenschaft. Akademische Verlagsgesellschaft, Frankfurt/Main. 33 rpm phonograph records.

Includes recordings made by Walther Gerlach, Werner Heisenberg, Max Planck; two others, by Max Born and Otto Hahn, are not presently available but may be reissued.

Science Jackdaws (see above, #17).

II

GUIDE TO ORIGINAL WORKS OF HISTORICAL IMPORTANCE AND THEIR TRANSLATIONS INTO OTHER LANGUAGES

FOREWORD

One of the recommendations of the International Working Seminar on the Role of the History of Physics in Physics Education, which was held at the Massachusetts Institute of Technology from 13 - 17 July 1970, is that the International Commission on Physics Education "cooperate with the Commission on Teaching of the International Union for History and Philosophy of Science and other scientific unions in establishing a committee on translations in the history of science." It is further suggested that "such a committee should act as a clearinghouse for exchange of views of science teachers and historians on the desirability of translations of particular books and articles of historical importance, and for exchange of information on plans for such translations with translators and publishers." Several participants in the seminar volunteered to serve on this committee under the chairmanship of Stephen Brush. This guide to original works of historical importance and their translations into other languages, with an appended list of works needing translation into English, is a preliminary report by the committee.

On behalf of the International Commission on Physics Education, the Organizing Committee, and the participants in the seminar, we gratefully acknowledge the financial support of the Alfred P. Sloan Foundation, the National Science Foundation, UNESCO, and the International Union of Pure and Applied Physics.

A.L. King
Chairman of the
Organizing Committee

January 1972

Part II

TABLE OF CONTENTS

1. INTRODUCTION

A Committee on Translations has been formed to set up an international clearinghouse for the exchange of information on translations of scientific works of historical importance. It will maintain a list of existing translations into various languages and translations in progress, and will solicit opinions about which works should be translated in the future. As the first concrete step toward this objective, we present below a list of books and articles in physics, indicating existing translations known to us.

All readers of this guide are urged to send information about existing translations of the works listed here, suggestions for other works to be added to the list, and opinions about the need for new translations of particular works. The latter may be submitted simply by sending in a list of names of authors with the item numbers given in the list, indicating language into which the works should be translated, and order of priority. For works not on the list it would be greatly appreciated if you could provide complete bibliographic information. Please keep in mind that this is <u>not</u> intended to be a comprehensive list of all translations of physics books and articles, but only those which are sufficiently important that they should be available to physics <u>students</u>.

The Committee was established originally under the sponsorship of the American Association for the Advancement of Science, Section L, and the International Union for History and Philosophy of Science, Commission on Teaching. The endorsement of the I.U.P.A.P. International Commission on Physics Education and of corresponding Commissions in the other international scientific unions is now being sought.

2. CLEARINGHOUSE ON TRANSLATIONS

Scientists of all times and all places seem to have felt a need for learning about what other scientists were doing and writing, hence there has always been a demand for translations. This was true even a few centuries ago when Latin was supposedly the universal language of the scholarly world; it is still true today although English may be fairly widely used for current scientific communication.

A special need for translations of scientific works arises in connection with teaching the history of science, or teaching science from a historical viewpoint. We are not so much concerned with the scholar doing research in the history of science, for he is expected to be able to read the original documents and should not trust anyone else's translation on a crucial point of technical detail or interpretation. But many teachers recognize the value of letting students, whose linguistic skills (at least in the U.S.A.) are grossly inadequate for reading scientific works in foreign languages, use selected excerpts from scientific works in courses in science as well as in the history of science. Teachers themselves, and graduate students wanting to acquire a broader knowledge of the literature of their specialty, can profit from the study of translations of longer and more technical works.

Translations of many of the older classics such as Newton's _Principia_ and Galileo's _Discorsi_ are easily available (though their accuracy has recently been questioned by historians of science). The science teacher, however, will often be more interested in the history of recent science, and here a special difficulty arises in "retrieving" the existing translations from the library. Almost all of the scientific discoveries of the past two centuries were announced in short articles in periodicals rather than in books. One cannot easily look up translations of these articles in a library book catalog. If a translation of the article was published, it may have been in another periodical or in a "sourcebook" which will be indexed under the editor's name. For a few of the major scientists, bibliographies have been published that list translations, but these lists quickly become obsolete as new translations appear.

The preliminary survey of the physics literature which forms the main part of this guide indicates that a large number of the most important articles and books are now available in English, and that before long an English version of almost any work important enough to be read by students will be in print somewhere. This should apply to other major languages. But the magnitude of the retrieval problem is shown by the fact that sometimes two or three independent translations of the same article have been published by translators unaware of each other's work. While it may in fact be desirable to retranslate a few major works such as those of Aristotle and Newton, it seems to us that translation is not such a profitable business as to justify so much duplication. We would prefer to have translators devote their energies to the more important works that are not yet available in all major languages. Moreover, by focusing on those works that teachers and historians of science think need to be translated, we hope to demonstrate a potential market for those translations that might otherwise be considered of too little interest to be published.

While one purpose of the committee is to encourage new translations for which a need is found to exist, we do not propose to sponsor such translations directly or to endorse any particular translation after it is published. Where possible we could bring together prospective translators, prospective publishers, and even prospective subsidizers of translations.* But it must be made clear that all information collected by the committee is to be publicly available to anyone who requests it.

We are starting with a list of works in physics only because of the special interests of the people who started this project. We urgently request the cooperation of teachers and scholars in other fields so that the list may cover all of the sciences.

The committee at present consists of the following people:

Stephen G. Brush (Chm.)	Institute for Fluid Dynamics & Applied Mathematics, University of Maryland, College Park, Md., USA [20740].
Richard Berendzen	Dept. of Astronomy, Boston University.
Harold L. Burstyn	Dept. of History, Carnegie-Mellon University, Pittsburgh.
Robert S. Cohen	Dept. of Physics, Boston University.
Samuel Devons	Dept. of Physics, Columbia University, New York.
Yehuda Elkana	Dept. of History and Philosophy of Science, The Hebrew University of Jerusalem.
Armin Hermann	Lehrstuhl für Geschichte der Naturwissenschaften und Technik, Stuttgart University.
Tetu Hirosige	Dept. of Physics, Nihon University, Tokyo.

*It should be noted that there already exist lists of potential translators, e.g., Frances E. Kaiser, ed., Translators and Translations: Services and Sources in Science and Technology. 2nd ed. New York: Special Libraries Association, 1965.

Michael Hoskin	Churchill College, Cambridge, England.
Max Jammer	Dept. of Physics, Bar-Ilan University, Ramat-Gan, Israel.
Giovanni Jona-Lasinio	Inst. of Physics, University of Rome.
Hans Kangro	Dept. of History of Science, Hamburg University.
Raymond J. Seeger	Ex officio Secretary, Section L, American Association for the Advancement of Science.
Roger H. Stuewer	School of Physics & Astronomy, University of Minnesota.
George W. White	Dept. of Geology, University of Illinois, Urbana.

3. PRELIMINARY LIST OF BOOKS AND ARTICLES IN PHYSICS

The list is arranged alphabetically by surnames of the first author, and numbered chronologically when there is more than one work by an author. There are cross-references for names of authors other than the first. The original title is given first, except when it was in Japanese, Greek, etc. Titles of translations are given when easily available; however, in several cases where a translation of an article appeared in another periodical, the citation was taken from a reference work such as the Royal Society Catalogue which does not give the title of the translation.

Perhaps the greatest source of uncertainty and error in such a list is the fact that many published translations are incomplete, yet this fact cannot be determined without detailed comparison of the translation and the original. It is our intention to indicate clearly which translations are incomplete, and to omit entirely those translations which do not include the major part of the original work; but so far it has not been possible to do the necessary checking.

In many cases we have listed reprints of the original work or of a translation, since this information was often available in the same source that was consulted for information about translations, and should also be useful to the same audience. However, this is not intended to be a complete index to reprints, and we have not tried to determine whether the reprints are still in print.

Franz Ulrich Theodor AEPINUS (1724-1802)

1. Tentamen Theoriae Electricitatis et Magnetismi. St. Petersburg, 1759. 390pp.

   Russian trans. by S. Ya. Lurye, ed. Ya. G. Dorfman: Teoriya elektrichestva i magnetizma (Leningrad: Izd-vo Akademii Nauk SSSR, 1951). (In the series "Klassiki Nauki.")

Jean Lerond d'Alembert (1717-1783)

1. Traité de dynamique, dans lequel les loix de l'équilibre & du mouvement des corps sont réduites au plus petit nombre possible, & démonstrées d'une manière nouvelle, & où l'on donne un principe général pour trouver le mouvement de plusieurs corps qui agissent les uns sur les autres d'une manière quelconque. Paris: Chez David l'aîne, 1743; new ed. 1758; reprinted by Gauthier Villars, Paris, 1921.

   German trans. by Arthur Korn, Abhandlung über Dynamik (Leipzig: Engelmann, 1899; Ostwalds Klassiker nr. 106).

   Russian trans. by V.P. Egorshin, Dinamika (Moscow: Gos. Izd.-vo Tekniko-Teoret. Lit-ry, 1950) in the series Klassiki Estestvoznaniya.

2. Recherches sur le courbe que forme une corde tendue mise en vibration. Mém. Acad. Sci. (1747), p. 214.

E. AMBLER

See C.S. WU.

Guillaume AMONTONS (1663-1705)

1. Moyen de substituer commodement l'action du feu a la force des hommes et des chevaux pour mouvoir les machines.

   Memoires de l'Academie Royale des Sciences, Paris, 1699, pp. 112-134.

2. Discours sur quelques proprietes de l'air, & le moyen d'en connoître la température dans tous les climats de la Terre.

   Memoires de l'Academie Royale des Sciences, Paris, 1702, pp. 155-174.

   English translation (partial), "An Air Pressure Thermometer" in Magie, Source Book in Physics. New York: McGraw-Hill, 1935; reprinted by Harvard University Press.

Andre Marie AMPÈRE (1775-1836)

1. Mêmoire présenté à l'Academie Royal des Sciences, le 2 Octobre 1820, où se trouve compris le résumé de ce qui avait été lu à la même Académie les 18 et 25 Septembre 1820, sur les effets des courans électriques.

   Ann. Chim. Phys., series 2, vol. 15 (1820), pp. 59-76, 170-218.

Partial English trans. by O.M. Blunn, "The Mutual Action of Two Electric Currents" in R.A.R. Tricker, Early Electrodynamics. Oxford: Pergamon Press, 1965.

2. Recueil d'Observations électrodynamiques. Paris, 1822. Reprinted under the title Memoires sur l'Électromagnetisme et l'Électrodynamique by Gauthier-Villars et Cie, Paris: 1921, in the series Les Maitres de la Pensée Scientifique.

   Includes: "De l'action exercée sur un courant électrique, par un autre courant, le globe terrestre ou un aimant"; and "Sur la détermination de la formule qui représente l'action mutuelle de deux portions infiniment petites de conducteurs voltaïques."

3. Mémoire sur la théorie mathématique des phénomènes électrodynamiques uniquement déduite de l'experience.

   Mem. Acad. Roy. Sci. 1823 (pub. 1827), vol. 6, pp. 175-387.

   English trans. by O.M. Blunn, "On the Mathematical Theory of Electrodynamic Phenomena, Experimentally Deduced" in R.A.R. Tricker, Early Electrodynamics. Oxford: Pergamon Press, 1965.

Carl D. ANDERSON (1905-)

1. The Positive Electron.

   Physical Review, series 2, vol. 43, pp. 491-494 (March 15, 1933). Reprinted in R.T. Beyer, Foundations of Nuclear Physics. New York: Dover Publications, Inc., 1949.

   Also reprinted in H.A. Boorse & L. Motz, The World of the Atom. New York: Basic Books, 1966.

ARCHIMEDES (287 ?-212 B.C.)

1. (On Floating Bodies——orig. in Greek.)

   English trans. "On Floating Bodies" in T.L. Heath, The Works of Archimedes. New York: Dover Publications (reprint of 1879 ed.).

   Heath trans. reprinted by Peter Wolff, Breakthroughs in Physics. New York: The New American Library, 1965; Signet Science Library Book paperback edition.

   French trans. by F. Peyrard, "Des corps qui sont portés sur un fluide" in Oeuvres d'Archimede (Paris, 1844).

   Another French trans. by Paul Ver Eecke, "Les corps flottant" in Les Oeuvres Complètes d'Archimede (Paris, 1920).

German trans. by Arthur Czwalina in Über schwimmende Körper und die Sandzahl. Leipzig: Akademische Verlagsgesellschaft m.b.h. 1925; Ostwalds Klassiker, nr. 213.

2. (On the Equilibrium of Planes, or the Centres of Gravity of Planes——orig. in Greek.)

   English trans. "On the Equilibrium of Planes" in T.L. Heath, The Works of Archimedes. New York: Dover Publications; reprint of 1879 ed.

   Heath trans. reprinted by Peter Wolff, Breakthroughs in Physics.

   French trans. by F. Peyrard, "De l'êquilibre des plans ou deleurs centres de gravitê" in Oeuvres d'Archimede (Paris, 1844).

   Another French trans. by Paul Ver Eecke, "De l'êquilibre des plans, ou Des centres de gravitê des plans" in Les Oeuvres Complètes d'Archimede (Paris, 1921).

   German trans. in Ostwalds Klassiker, nr. 203. Leipzig: Akad. Verlags., 1923.

ARISTOTLE (384-322 B.C.)

1. (Physics——orig. in Greek.)

   English translations——several are available and can easily be located in library card catalogs. Two recent ones are available in paperback: Physics by H.G. Apostle (Bloomington: Indiana Uiversity Press, 1969); and Aristotle's Physics by W. Charlton (New York & London: Oxford University Press, 1970).

   German trans. by Olof Gignon, Physik, in Aristoteles Werk (Zürich: Artemis-Verlag, 1950), Bd. 6; also by Paul Gohlke, in Die Lehrschriften, 4.1 (Paderborn, 1956).

   French trans. by J. Barthelmy-Saint-Hilaire, Physique, in Oeuvres d'Aristote (Paris, 1862), vols. xii-xiii.

Amedeo AVOGADRO (1776-1856)

Essai d'une manière de determiner les masses relatives des molêcules êlêmentaires des corps, et les proportions selon lesquelles elles entrent dans ces combinations.

Journal de Physique, vol. 73 (1811), pp. 58-76.

English trans., "Essay on a Manner of Determining the Relative Masses of the Elementary Molecules of Bodies, and the Proportions in which They Enter into These Compounds" in Foundations of the Molecular Theory, Alembic Club Reprints No. 4 (re-issue ed., Edinburgh: E. & S. Livingstone Ltd., 1961).

Johann Jakob BALMER (1825-1898)

1. Notiz über die Spectrallinien des Wasserstoffs.

   Verhandlungen der Naturforschenden Gesellschaft zu Basel, vol. 7, pp. 548-560, 750-752, reprinted in Ann. Physik, series 3, vol. 25 (1885), pp. 80-87.

   English translation by J.B. Sykes, "A Note on the Spectral Lines of Hydrogen" in W.R. Hindmarsh, Atomic Spectra. Oxford: Pergamon Press, 1967.

John BARDEEN, L.N. COOPER and J.R. SCHRIEFFER

Theory of Superconductivity.

Phys. Rev., series 2, vol. 108 (1957), pp. 1175-1204.

[Antoine-] Henri BECQUEREL (1852-1908)

1. Sur les radiations émises par phosphorescence.

   Compt. rend. Acad. Sci. Paris, vol. 122 (1896), pp. 420-421.

   English trans. "On the Radiation Emitted in Phosphorescence" in Alfred Romer, The Discovery of Radioactivity and Transmutation. New York: Dover Publications, 1964.

2. Sur les radiations invisible émises par les corps phosphorescents.

   Compt. rend. Acad. Sci. Paris, vol. 122 (1896), pp. 501-503.

   English translation "On the Invisible Radiations Emitted by Phosphorescent Substances" in Alfred Romer, The Discovery of Radioactivity and Transmutation.

   Sur les radiations invisible émises par les sels d'uranium.

   Compt. rend. Acad. Sci. Paris, vol. 122 (1896), pp. 689-694.

   English trans. "On the Invisible Radiations Emitted by the Salts of Uranium" in Alfred Romer, The Discovery of Radioactivity and Transmutation.

Daniel BERNOULLI (1700-1782)

1. Hydrodynamica, sive de viribus et motibus fluidorum commentarii. Argentorati (Strassburg): Johann Reinhold Dulsecker, 1738.

English translation: Hydrodynamics, by Thomas Carmody and Helmut Kobus, with preface by Hunter Rouse. New York: Dover Publications, Inc., 1968. (Bound with translation of Johann Bernoulli's Hydraulica.)

German trans. by Karl Flierl, in: Veröffentlichungen des Forschungsinstituts des Deutschen Museum für die Geschichte der Naturwissenschaften und der Technik. Reihe C: Quellentexte und Übersetzungen, Nr. 1a & 1b (München, 1963).

2. Reflexions et eclaircissements sur les nouvelles vibrations des cordes, exposes dans les memoirs de l'Academie, de 1747 et 1748 (Berlin: 1775), pp. 147-

Jean or Johann BERNOULLI (1667-1748)

1. Hydraulica, Nunc primum detecta ac demonstrata directe ex fundamentis pure mechanicis, Anno 1732. (Published in the Memoirs of the Imperial Academy of Science, St. Petersburg, for 1737 and 1738, printed in 1744 and 1747 resp. The date 1732 which appears on the title page is now considered to be false.)

   English translation by Thomas Carmody and Helmut Kobus, Hydraulics; Preface by Hunter Rouse. New York: Dover Publications, Inc., 1968. (Bound with translation of Daniel Bernoulli's Hydrodynamica.)

Hans BETHE (1906-)

1. Termaufspaltung in Kristallen.

   Annalen der Physik, series 5, vol. 3 (1929), pp. 133-208.

   English translation "Splitting of Terms in Crystals," published by Consultants Bureau, New York.

   Partial English translation in Arthur P. Cracknell, Applied Group Theory. Oxford: Pergamon Press, 1968.

Jean Baptiste BIOT (1774-1862) and Felix SAVART

1. Note sur le magnetisme de la Pile de Volta.

   Ann. Chim. Phys., series 2, vol. 15 (1820), pp. 222-223.

   English trans. by O.M. Blunn in R.A.R. Tricker, Early Electrodynamics. Oxford: Pergamon Press, 1965.

   German trans. in Ann. Physik, vol. 66 (1820), pp. 392-394.

George David BIRKHOFF (1884-1944)

1. Proof of the Ergodic Theorem.

   Proc. Nat. Acad. Sci. USA, Vol. 17, pp. 656-660 (1931). Reprinted in R. Bellman (ed.), A Collection of Modern Mathematical Classics: Analysis. New York: Dover, 1961.

Joseph BLACK (1728-1799)

1. Lectures on the Elements of Chemistry. Ed., J. Robinson. Edinburgh, 1803.

   German trans. by Crell: Vorlesungen über der Grundlehren der Chemie. Hamburg, 1804.

Nikolai Nikolaevich BOGOLIUBOV (1909-)
or BOGOLYUBOV
or BOGOLUBOV

1. Problemy dinamischeskoi teori v statisticheskoi fiziki. Moscow: Gos. Izd. Tekh.-teor. Lit., 1946.

   English translation by E.K. Gora: "Problems of a Dynamical Theory in Statistical Physics," in J. de Boer and G.E. Uhlenbeck, Studies in Statistical Mechanics, Vol. 1. Amsterdam: North-Holland Pub. Co., 1962.

Nikolai Nikolaevich BOGOLIUBOV, B.V. MEDVEDEV
and M.K. POLIVANOV

Voprosy teorii dispersionnykh Sootnoshenii. Moscow: Gosudarstvennoe Izdatel'stvo Fiziko-Matematicheskoi Literatury, 1958.

English trans. by S.G. Brush. Theory of Dispersion Relations, University of California, Lawrence Radiation Laboratory. Translation UCRL Trans. 499. (L)

Niels BOHR (1885-1962)

1. On the Constitution of Atoms and Molecules.

   Philosophical Magazine, Ser. 6, Vol. 26, pp. 1-25, 476-502, 857-875 (1913).

   Reprinted in Niels Bohr: On the Constitution of Atoms and Molecules (papers of 1913 reprinted from the Phil. Mag. with an introduction by L. Rosenfeld). Copenhagen: Munksgaard; and New York: Benjamin, 1963.

   Reprinted (in part) in D. ter Haar, The Old Quantum Theory. Oxford: Pergamon Press, 1967.

Reprinted in W.R. Hindmarsh, Atomic Spectra. Oxford: Pergamon Press, 1967.

German translation in N. Bohr, Abhandlungen über Atombau aus den Jahren 1913-1916; trans. by H. Stintzing (Braunschweig: F. Vieweg, 1921; reprinted in Dokumente der Naturwissenschaft-Abteilung Physik. Stuttgart: Ernst Battenberg Verlag, 1964. Vol. 5, pp. 33-57, 58-83, 84-101.

2. On the Quantum Theory of Line-Spectra.

Mem. Acad. Roy. Sci. & Lett. Copenhagen (D. Kgl. Danske Vidensk. Selsk. Skrifter, Series 8, Vol. 4, Fasc. 1-3, 1917).

First part only reprinted in B.L. van der Waerden. Sources of Quantum Mechanics. Amsterdam: North-Holland Pub. Co., 1967; New York: Dover Publications.

German trans. by P. Hertz, Über die Quantentheorie der Linienspektren. Braunschweig: Vieweg, 1923.

Niels BOHR, H.A. KRAMERS and J.C. SLATER

3. The Quantum Theory of Radiation.

Phil. Mag., Series 6, Vol. 47, pp. 785-802 (1924).

Reprinted in B.L. van der Waerden. Sources of Quantum Mechanics. Amsterdam: North-Holland Pub. Co., 1967; New York: Dover Publications.

German trans. "Über die Quantentheorie der Strahlung." Zeits. f. Physik, Vol. 24, pp. 69-87 (1924).

Niels BOHR

4. Discussion with Einstein on Epistemological Problems in Atomic Physics, pp. 201-204, in P.A. Schilpp, ed., Albert Einstein: Philosopher-Scientist. Library of Living Philosophers, 1949; reprinted by Harper Torchbooks, 1959.

See also A. Einstein.

Ludwig BOLTZMANN (1844-1906)

1. Weitere Studien über das Wärmegleichgewicht unter Gasmolekülen.

Sitzungsberichte Akad. Wiss. Wien. Vol. 66 (Abt. II), pp. 275-370 (1872).

Reprinted in Boltzmann's Wissenschaftliche Abhandlungen. Leipzig: J.A. Barth, 1909; New York: Chelsea, 1968.

English translation "Further Studies on the Thermal Equilibrium of Gas Molecules," in S.G. Brush, Kinetic Theory, Vol. 2. Oxford: Pergamon Press, 1966.

2. Ableitung des Stefan'schen Gesetzes betreffend die Abhängigkeit der Wärmestrahlung von der Temperatur aus der electromagnetischen Lichttheorie.

   Ann. Physik, Series 3, Vol. 22, pp. 291-294 (1884).

   Reprinted in his Wissenschaftliche Abhandlungen.

3. Vorlesungen über Gastheorie.

   I. Teil. Leipzig: J.A. Barth, 1896.

   II. Teil. Leipzig: J.A. Barth, 1898.

   French translation by A. Gallotti and H. Benard, with an introduction and notes by M. Brillouin. Leçons sur la theorie des gaz. Paris: Gauthier-Villars, 1902 and 1905.

   Russian translation, edited with introduction and notes by B.I. Davydov. Lektsii po teori gasov. Moskva: Gosurdarstvennoe izdatelstvo tekhniko-teoreticheckoi literatury, 1956.

   English translation, with introduction and notes by Stephen G. Brush. Lectures on Gas Theory. Berkeley: University of California Press, 1964.

Max BORN (1882-1970) and Theodore VON KÁRMÁN

1. Über Schwingungen in Raumgittern.

   Physik. Zeits. Vol. 13, pp. 297-309 (1912).

   Reprinted in Born's Ausgewählte Abhandlungen. Göttingen: Van den Hoeck & Ruprecht, 1963. Vol. 1.

2. Zur Theorie der spezifischen Wärme.

   Physik. Zeits. Vol. 14, pp. 15-19 (1913).

   Reprinted in Born's Ausgewählte Abhandlungen.

3. Über die Verteilung der Eigenschwingungen von Punktgittern.

   Physik. Zeits. Vol. 14, pp. 65-71 (1913).

   Reprinted in Born's Ausgewählte Abhandlungen.

Max BORN

4. Kritische Betrachtungen zur traditionellen Darstellung der Thermodynamik.

Physik. Zeits. Vol. 22, pp. 218-224, 249-254, 282-286 (1921).

Reprinted in his Ausgewählte Abhandlungen. Göttingen: Van den Hoek & Ruprecht, 1963. Vol. 1

Max BORN and Pascual JORDAN

5. Zur Quantenmechanik.

   Zeits. Physik, Vol. 34, pp. 858-888 (1925).

   Reprinted in Born's Ausgewählte Abhandlungen.

   English translation "Quantum Mechanics" in G. Ludwig, Wave Mechanics. Oxford: Pergamon Press, 1968 (portions omitted).

   Another English translation, omitting Chapter 1 but including some of the portions omitted from the Ludwig translation, is in B.L. van der Waerden. Sources of Quantum Mechanics. Amsterdam: North-Holland Pub. Co., 1967; reprinted by Dover.

Max BORN

6. Das adiabatenprinzip in der Quantenmechanik.

   Zeits. f. Physik, Vol. 40, pp. 167-192 (1926).

   Reprinted in Ausgewählte Abhandlungen.

7. Physical Aspects of Quantum Mechanics.

   Nature, Vol. 119, pp. 354-357 (1926). Read at the Oxford Meeting of the British Association for the Advancement of Science, August 10, 1926; translated into English by R. Oppenheimer.

8. Quantenmechanik der Stossvorgänge.

   Zeits. Physik, Vol. 38, pp. 803-827 (1926).

   Reprinted in his Ausgewählte Abhandlungen.

   English translation "Quantum Mechanics of Collision Processes" in Gunter Ludwig, Wave Mechanics. Oxford: Pergamon Press, 1968.

Max BORN, Werner HEISENBERG, and Pascual JORDAN

9. Zur Quantenmechanik II.

   Zeits. Physik, Vol. 35, pp. 557-615 (1926).

   Reprinted in Born's Ausgewählte Abhandlungen.

English translation "On Quantum Mechanics II" in B.L. van der Waerden, Sources of Quantum Mechanics. Amsterdam: North-Holland Pub. Co., 1967; reprinted by Dover.

Max BORN and Joseph E. MAYER

10. Zur Gittertheorie der Ionenkristalle.

    Zeits f. Physik, Vol. 75, pp. 1-18 (1932).

    Reprinted in Born's Ausgewählte Abhandlugen.

Max BORN

11. Bemerkungen zur statistischen Deutung der Quantenmechanik in Werner Heisenberg und die Physik unserer Zeit. Braunschweig: Vieweg, 1961; pp. 103-118.

    Reprinted in his Ausgewählte Abhandlungen.

Rudjer Josip BOŠKOVIČ
or Roger Joseph BOSCOVICH (1711-1787)

1. Theoria Philosophiae Naturalis redacta ad unicam legem virium in natura existentium.

   Venetian edition of 1763 (revised and enlarged by the author from the first edition, Vienna, 1758).

   Translated by J.M. Child, A Theory of Natural Philosophy. Open Court Publishing Co., Chicago, 1922 ? (includes Latin edition on facing pages).

   Reprinted (without the Latin edition) by M.I.T. Press, Cambridge, 1966.

Satyendranath BOSE (1894-)

1. Plancks Gesetz und Lichtquantenhypothese.

   Zeitschrift für Physik, Vol. 26, pp. 178-181 (1924). (This is a German translation by A. Einstein; the original English manuscript was never published.)

   Reprinted in K.B. Beaton and H.C. Bolton, A German Source-Book in Physics. Oxford University Press, 1969.

   English translation in H.A. Boorse & L. Motz. The World of the Atom. New York: Basic Books, 1966.

Robert BOYLE (1627-1691)

1. *New experiments physico-mechanicall, touching the spring of the air, and its effects; made, for the most part in a new pneumatical engine. Written by way of Letter to the right Honourable Charles Lord Viscount of Dungarvan, eldest son to the Earl of Corke*. Oxford, 1660.

   Reprinted in Boyle's *Works*. Ed., T. Birch. London: 2nd ed., 1772.

2. *A Defence of the doctrine touching the spring and weight of the air, proposed by Mr. R. Boyle in his New Physico-Mechanical Experiments; Against the objections of Franciscus Linus, wherewith the Objector's Funicular Hypothesis is also examined*. Oxford, 1662.

   Reprinted in Boyle's *Works*. Ed., T. Birch. London: 1744; 2nd ed., 1772.

Louis-François-Clement BREGUET (1804-1883)

See A.H.L. FIZEAU.

Gregory BREIT (1899-) and Eugene WIGNER

1. Capture of Slow Neutrons.

   Physical Review, Series 2, Vol. 49, pp. 519-531 (1936).

   Reprinted in I.E. McCarthy. *Nuclear Reactions*. Oxford: Pergamon Press, 1970.

Louis, Prince de BROGLIE (1892-)

1. *Recherches sur la theorie des quanta*.

   (Thèse, Faculté des Sciences de l'Université de Paris) published by Masson et Cie., Paris, 1924; also in Annales de Physique, Series 10, Vol. 3, pp. 22-128. 1925.

   Partial English translation in G. Ludwig. *Wave Mechanics*. Oxford: Pergamon Press, 1968, pp. 73-93.

   German trans. by Walther Becker. *Untersuchungen zur Quantentheorie*. Leipzig: Akademische Verlagsgesellschaft m.b.h., 1927.

Robert BROWN (1773-1858)

1. *A brief account of microscopical observations made in the months of June, July and August, 1827, on the particles contained in the pollen of plants; and on the general existence of active molecules in organic and inorganic bodies*.

Printed for private circulation, 1828; reprinted in the Edinburgh New Philos. J., Vol. 5, pp. 358-371 (1828); reprinted in Phil. Mag., Vol. 4, pp. 161-173 (1828); reprinted in The Miscellaneous Botanical Works of Robert Brown, published for the Ray Society by R. Hardwicke (1866).

French trans. in Annales des Sciences Naturelles (Paris), Vol. 14, pp. 341-362 (1828).

German trans. in Froriep's Notizen aus dem Gebiete der Natur- und Heilkunde, Vol. 22, pp. 161-170 (1828); Oken's Isis, Vol. 21, pp. 1006-1012 (1828); Annalen der Physik, Vol. 90, pp. 294-313 (1828).

See also German trans. with extensive historical commentary by F.I.F. Meyer, in Robert Brown's Vermischte Botanische Schriften. Ed. C.G. Nees von Esenbeck. Leipzig: 1825-1834. Vol. 4

Constantin CARATHÉODORY (1873-1950)

1. Untersuchungen über die Grundlagen der Thermodynamik.

   Mathematische Annalen Vol. 67, pp. 355-386 (1909).

   English translation "Investigations of the Fundamentals of Thermodynamics." Translation number 1574, SIA Translation Center, the John Crerar Library, 86 East Randolph Street, Chicago, Illinois. (SGB was informed that this translation is "damaged—no longer available").

   Another English translation by T.E. Phipps, Donald Miller and Michael Fitzel. "Investigations of the Foundations of Thermodynamics." Unpublished. (SGB has copy.)

2. Über den Wiederkehrsatz von Poincaré.

   Situngsberichte der Preussischen Akademie der Wissenchaften, 10 July 1919, pp. 579-584.

   Reprinted in his Gesammelte Mathematische Schriften. München: Beck, 1954-57.

   English translation by Stephen G. Brush. "On Poincaré's Recurrence Theorem." University of California, Lawrence Radiation Laboratory, Livermore, California. Report UCRL Trans-871(L).

3. Über die Bestimmung der Energie und der absoluten Temperatur mit Hilfe von reversiblem Prozessen.

   Sitzungsberichte der Preussischen Akademie der Wissenschaften, 29 January 1925, pp. 39-47.

Nicolas Leonard Sadi CARNOT (1796-1832)

1. Rêflexions sur la puissance motrice du feu et sur les machines propres a developper cette puissance. Paris, 1824.

   English translation by R.H. Thurston: Reflections on the Motive Power of Heat. New York: Macmillan, 1890.

   Thurston's translation reprinted with introduction, notes, and an appendix. "Selections from the Posthumous Manuscripts of Carnot" by E. Mendoza, in Reflections on the Motive Power of Fire by Sadi Carnot and Other Papers on the Second Law of Thermodynamics by E. Clapeyron and R. Clausius. New York: Dover Publications, 1960.

   German trans. by W. Ostwald. Betrachtungen über die bewegende Kraft des Feuers und die zur Entwicklung diesen Kraft geeigneten Maschinen. Leipzig: W. Engelmann, 1892; Ostwald's Klassiker nr. 37.

Henry CAVENDISH (1731-1810)

1. The Scientific Papers of the Honourable Henry Cavendish. Volume I. The electrical researches of the Honourable Henry Cavendish, F.R.S., written between 1771 and 1781, edited from the original manuscripts in the possession of the Duke of Devonshire, K.G., by J. Clerk Maxwell, F.R.S. Cambridge University Press, 1879.

   Volume II. Chemical and Dynamical, ed. by Sir E.Thorpe. Cambridge University Press, 1921.

James CHADWICK (1891-   )

1. Possible existence of a neutron.

   Nature, Vol. 129, p. 312 (1932).

   Reprinted in Stephen Wright. Classical Scientific Papers-Physics. New York: American Elsevier Pub. Co., 1965.

2. The Existence of a Neutron.

   Proceedings of the Royal Society of London, Vol. A136, pp. 692-708 (1932).

   Reprinted in R.T. Beyer, Foundations of Nuclear Physics. New York: Dover Publications, Inc., 1949; also in Wright, Classical Scientific Papers-Physics.

   Reprinted (with two or three paragraphs omitted) in H.A. Boorse & Lloyd Motz, The World of the Atom. New York: Basic Books, 1966.

Subrahmanyan CHANDRASEKHAR (1910-)

1. Stochastic Problems in Physics and Astronomy.

   Reviews of Modern Physics, Vol. 15, pp. 1-89 (1943).

   Reprinted in Nelson Wax. Selected Papers on Noise and Stochastic Processes. New York: Dover Publications, 1954.

Sydney CHAPMAN (1888-1970)

1. On the Law of Distribution of Molecular Velocities, and on the Theory of Viscosity and Thermal Conduction, in a Non-Uniform Simple Monatomic Gas.

   Phil. Trans. Roy. Soc. London, Vol. A 216, pp. 279-348 (1916).

2. On the Kinetic Theory of a Gas; Part II, A Composite Monatomic Gas, Diffusion, Viscosity, and Thermal Conduction.

   Phil. Trans. Roy. Soc. London, Vol. A 217, pp. 115-197 (1917).

3. A Note on Thermal Diffusion. (Co-author, F.W. Dootson)

   Philosophical Magazine, Series 6, Vol. 33, pp. 248-253 (1917).

   Reprinted in S.G. Brush. Kinetic Theory. Vol. 3. Oxford: Pergamon Press, 1972.

4. The Mathematical Theory of Non-Uniform Gases. An Account of the Kinetic Theory of Viscosity, Thermal Conduction, and Diffusion in Gases.

   Cambridge University Press, 1939; 2nd ed. 1952; 3rd ed. 1970.

Ernst Florens Friedrich CHLADNI (1756-1827)

1. Entdeckungen über die Theorie des Klanges. Leipzig: Weidmann, Erben & Reich, 1787.

[Benoit-Paul-] Emile CLAPEYRON (1799-1864)

1. Mémoire sur la puissance motrice de la chaleur.

   J. Ecole Polyt. Paris, Vol. 14, Cahier 23, pp. 153-190 (1834).

German translation in Annalen der Physik, Ser. 2, Vol. 59, pp. 446-450 (1834).

English trans. "Memoir on the Motive Power of Heat" in Richard Taylor, ed., Scientific Memoirs, Vol. I. London, 1837; Johnson Reprint Corp., New York, 1966.

Another English translation, "Memoir on the Motive Power of Heat" by E. Mendoza, in Mendoza's book, Reflections on the Motive Power of Fire by Sadi Carnot and Other Papers on the Second Law of Thermodynamics by E. Clapeyron and R. Clausius. New York: Dover Publications, 1960.

Rudolf Julius Emmanuel CLAUSIUS (1822-1888)

1. Ueber die bewegende Kraft der Wärme und die Gesetze welche sich daraus für die Wärmelehre selbst ableiten lassen.

   Annalen der Physik, Ser. 2, Vol. 79, pp. 368-397; 500-524 (1850).

   Reprinted with other writings on thermodynamics in his Abhandlungen über die mechanische Wärmetheorie, Part I. Vieweg: Braunschweig, 1864.

   English translation in Phil. Mag., Ser. 4, Vol. 2, pp. 1-21, 102-119 (1851).

   English translation "On the Motive Power of Heat, and on the Laws Which Can Be Deduced From It for the Theory of Heat" by W.F. Magie.

   Magie's translation reprinted by E. Mendoza in Reflections on the Motive Power of Fire by Sadi Carnot and Other Papers on the Second Law of Thermodynamics by E. Clapeyron and R. Clausius. New York: Dover Publications, 1960.

   See also the English translation of Clausius' Abhandlungen . . . , The Mechanical Theory of Heat. Ed. by T. Archer Hirst with an introduction by J. Tyndall. London: J. van Voorst, 1867; another translation, same title, by W.R. Browne. London: Macmillan, 1879.

   French trans. (partial ?) in Ann. Chim., Series 3, Vol. 35, pp. 482-503 (1852). Also Bibl. Univ. Archives (Genève), Vol. 36, pp. 194-206 (1857).

2. Ueber die Art der Bewegung, welche wir Wärme nennen. Annalen der Physik, Ser. 2, Vol. 100, pp. 353-380 (1857).

   English translation, "The Nature of the Motion Which We Call Heat" in Philosophical Magazine, Ser. 4, Vol. 14, pp. 108-127 (1857).

English translation reprinted in S.G. Brush, *Kinetic Theory*, Vol. 1. Oxford: Pergamon Press, 1965.

French trans. in Ann. Chim., Vol. 50, pp. 497-507 (1857) and in Bibl. Univ. Archives (Genève), Vol. 36, pp. 293-309 (1857).

3. Ueber die mittlere Länge der Wege, welche bei der Molecularbewegung gasförmigen Körper von den einzelnen Molecülen zurückgelegt werden; nebst einigen anderen Bemerkungen über die mechanische Wärmetheorie.

   Annalen der Physik, Ser. 2, Vol. 105, pp. 239-258 (1858).

   English translation by F. Guthrie, "On the Mean Lengths of the Paths Described by the Separate Molecules of Gaseous Bodies."

   Philosophical Magazine, Ser. 4, Vol. 17, pp. 81-91 (1859).

   Guthrie's translation reprinted in S.G. Brush, *Kinetic Theory*. Vol. 1. Oxford: Pergamon Press, 1965.

   French trans. in Bibl. Univ. Archives (Genève), Vol. 17, pp. 81-91 (1859).

4. Ueber verschiedene für die Anwendung bequeme Formen der Hauptgleichungen der mechanischen Wärmetheorie.

   Annalen der Physik, Ser. 2, Vol. 125, pp. 353-400 (1865).

   Reprinted in his *Abhandlungen über die mechanische Wärmetheorie*. Braunschweig, 1864.

   English translations in the two English editions of this book.

   "On several convenient forms of the fundamental equations of the mechanical theory of heat" in *The Mechanical Theory of Heat*, ed. by T. Archer Hirst. London: J. Van Voorst, 1867. And ditto by W.R. Browne, London: Macmillan, 1879.

5. Ueber einen auf die Wärme anwendbaren mechanischen Satz.

   Sitzungsberichte der Niederrheinischen Gesellschaft, Bonn, pp. 114-119 (1870).

   Reprinted in Ann. Physik, Series 2, Vol. 141, pp. 124-130 (1870).

   English translation: "On a Mechanical Theorem Applicable to Heat," *Philosophical Magazine*, Ser. 4, Vol. 40, pp. 122-127 (1870).

   English translation reprinted in S.G. Brush, *Kinetic Theory*, Vol. 1. Oxford: Pergamon Press, 1965.

William CLEGHORN

1. De Igne.

   Dissertation, 1779.

   English trans. by Douglas McKie and N.H. deV. Heathcote. "William Cleghorn's 'De Igne' (1779." Annals of Science, Vol. 14, pp. 1-82 (1958).

John Douglas COCKCROFT (1897-) and E.T.S. WALTON

1. Experiments with high velocity positive ions. II - The Disintegration of Elements by High Velocity Protons.

   Proc. Roy. Soc. London, Vol. A 137, pp. 229-242 (1932).

   Reprinted in Stephen Wright. Classical Scientific Papers--Physics. New York: American Elsevier Pub. Co., 1965.

Arthur Holly COMPTON (1892-1962)

1. A Quantum Theory of the Scattering of X-Rays by Light Elements.

   Phys. Rev., Series 2, Vol. 21, pp. 483-502 (1923).

2. X-Ray Spectra from a Ruled Reflection Grating. (Co-author R.L. Doan)

   Proc. Natl. Acad. Sci. USA. Vol. 11, pp. 598-601 (1925).

   Reprinted in Stephen Wright. Classical Scientific Papers --Physics. New York: American Elsevier Pub. Co., 1965.

Karl Taylor COMPTON (1887-1954)

See O.W. RICHARDSON.

Leon N. COOPER (1930-)

See J. BARDEEN.

Charles Augustin de COULOMB (1736-1806)

1. Sur l'électricité et le magnétisme.

   Hist. et Mem. Acad. Sci. Paris, 1785, pp. 569-638; 1786, pp. 67-77.

   German trans. by Walter König. Vier Abhandlung über die Elektricität und den Magnetismus. Leipzig: W. Engelmann, 1890; Ostwalds Klassiker, nr. 13.

Clyde Lorrain COWAN, Jr. (1919-)

see F. REINES.

Marie Sklodowska CURIE (1867-1934)

1. Rayons émis par les composés de l'uranium et du thorium.

   Compt. Rend. Acad. Sci. Paris, Vol. 126, pp. 1101-1103 (1898).

   English trans. "Rays Emitted by the Compounds of Uranium and Thorium" in Alfred Romer. Radiochemistry and the Discovery of Isotopes. New York: Dover Publications, 1970.

See also Pierre CURIE.

Pierre CURIE (1859-1906), Marie CURIE, and G. BÉMONT

1. Sur une nouvelle substance fortement radioactive, contenue dans la pechblende.

   Compt. Rend. Acad. Sci., Paris, Vol. 127, pp. 1215-1217 (1898).

   English translation. "On a New Substance, Strongly Radioactive, Contained in Pitchblende" in Henry A. Boorse & Lloyd Motz. The World of the Atom. New York: Basic Books, 1966.

   Another English trans. in Alfred Romer, Radiochemistry...

Pierre CURIE and Marie CURIE

2. Sur les corps radio-actifs.

   Compt. Rend. Acad. Sci. Paris, Vol. 134, pp. 85-87 (1902).

   English trans. "On Radioactive Substances" in Alfred Romer. The Discovery of Radioactivity and Transmutation. New York: Dover Publications, 1964.

John DALTON (1766-1844)

1. Experimental essays on the constitution of mixed gases: on the force of steam or vapour from water and other liquids in different temperatures, both in a Torricellian vacuum and in air: on evaporation: and on the expansion of elastic fluids by heat, etc.

   Manchester: R. and W. Dean and Co., 1802.

Also in Memoirs of the Manchester Lit. and Phil. Soc., Vol. 5, Part II, First Series, pp. 535-602 (1802).

French trans. in Ann. Chim., Vol. 46, pp. 250-276 (1803).

German trans. in Ann. Phys., Vol. 12, pp. 310-318, 385-395; Vol. 13, pp. 438-445 (1803).

2. On the Absorption of Gases by Water and Other Liquids.

   Memoirs of the Manchester Lit. and Phil. Society, Vol.6, pp. 271-287 (1805, read 21 Oct. 1803).

   Phil. Mag., Vol. 24, pp. 15-24 (1806).

   German trans. in Ann. Phys., Vol. 28, pp. 397-416 (1808).

   French trans in J. Phys., Vol. 65, pp. 57-68 (1807).

3. A New System of Chemical Philosophy. 2 vols. in 3.

   Manchester, printed by S. Russell for R. Bickerstaff. London, 1808-1827. 2nd ed., Part 1 only, London: John Weale, printed by Simpson and Gillett, Manchester, 1842.

   German translation by Friedrich Wolff: Ein neues System des chemischen Theiles der Naturwissenschaft. Berlin: Hitzig, 1812-1813, 2 pts. in 1 vol.

Clinton Joseph DAVISSON (1881-1958) and L.H. GERMER

1. Diffraction of Electrons by a Crystal of Nickel.

   Phys. Rev., Series 2, Vol. 30, pp. 705-740 (1927).

Peter Joseph Wilhelm DEBYE (1884-1966)

1. Einige Resultate einer kinetischen Theorie der Isolatoren.

   Phys. Zeits., Vol. 13, pp. 97-100 (1912).

   English translation, "Some Results of a Kinetic Theory of Insulators" in The Collected Papers of Peter J.W. Debye. New York & London: Interscience, 1954.

2. Zur Theorie der spezifischen Wärmen.

   Ann. Physik, Ser. 4, Vol. 39, pp. 789-838 (1912).

   English translation in his Collected Papers.

3. Interferenz von Röntgenstrahlen und Wärmebewegung.

   Ann. Physik, Ser. 4, Vol. 43, pp. 49-95 (1914).

   English translation, "X-Ray Interference and Thermal Movement" in his Collected Papers.

Peter J.W. DEBYE and P. SCHERRER

4. Interferenzen an regellos orientierten Teilchen im Röntgenlicht. I.

   Phys. Zeits., Vol. 17, pp. 277-283 (1916).

   English translation, X-Ray Interference Patterns of Particles Oriented at Random, I in his <u>Collected Papers</u>.

Peter J.W. DEBYE

5. Die van der Waalsschen Köhasionskräfte.

   Phys. Zeits., Vol. 21, pp. 178-187 (1920).

   English translation "Van der Waals' Cohesive Forces" in his <u>Collected Papers</u>.

6. Zerstreuung von Röntgenstrahlen und Quantentheorie.

   Phys. Zeits., Vol. 24, pp. 161-166 (1923).

   English trans. "X-Ray Scattering and Quantum Theory" in his <u>Collected Papers</u>.

Peter J.W. DEBYE and E. HÜCKEL

7. Zur Theorie der Elektrolyte. I. Gefrierpunktserniedrigung und verwandte Erscheinungen. II. Das Grenzgesetz für die elektrische Leitfähigkeit.

   Phys. Zeits., Vol. 24, pp. 185-206, 305-325 (1923).

   English translation, "On the Theory of Electrolytes. I. Freezing Point Depression and Related Phenomena. II. Limiting Law for Electric Conductivity" in <u>The Collected Papers of Peter J.W. Debye</u>.

Peter J.W. DEBYE and H. MENKE

8. Bestimmung der inneren Struktur von Flüssigkeiten mit Röntgenstrahlen.

   Phys. Zeits., Vol. 31, pp. 797-798 (1930).

   English translation, "Determination of the Inner Structure of Liquids by X-Rays" in <u>The Collected Papers of Peter J.W. Debye</u>.

René DESCARTES (1596-1650)

1. <u>Discours de la Méthode, pour bien conduire sa raison, et chercher la vérité dans les sciences, plus la dioptrique, Les Météores, et la geometriè qui sont des essais de cette méthode</u>. Leyden, 1937.

English and German translations of the Discourse on Method are available in several editions.

Complete translation of all four parts of this book by Paul J. Olscamp: Discourse on Method, Optics, Geometry and Meteorology. Indianapolis: Bobbs-Merrill Co., Inc., 1965.

2. Principia Philosophiae. Amsterdam, 1644. Reprinted in Oeuvres de Descartes. Paris, 1904.

   French translation, Les principes de la philosophie de René Descartes. Paris, 1647.

   4th edition, Paris, 1681.

   Reprinted in Oeuvres de Descartes, publiées par Charles Adam & Paul Tannery, Tome IX. Paris, 1904.

   Partial English translation by Elizabeth S. Haldane and G.R.T. Ross in The Philosophical Works of Descartes. Cambridge University Press, 1911. (The translation omits the sections dealing with the laws of collisions; Part II, paragraphs 26ff., and other parts relevant to physics.)

   English translation of Part II, paragraphs 24-40 in Marie Boas Hall, ed. Nature and Nature's Laws. Documents of the Scientific Revolution. New York: Harper & Row, 1970. (P)

   English trans. by Edward J. Collins, University of Chicago, in preparation.

   German translation by A. Buchenau. Die Prinzipien der Philosophie. Hamburg: Meiner, 1955.

Paul Adrien Maurice DIRAC (1902-)

1. The Fundamental Equations of Quantum Mechanics.

   Proc. Roy. Soc. London, Vol. A 109, pp. 642-653 (1926).

   Reprinted in B.L. Van der Waerden. Sources of Quantum Mechanics. Amsterdam: North-Holland Pub. Co., 1967; New York: Dover Publications.

2. Quantum Mechanics and Preliminary Investigation of the Hydrogen Atom.

   Proceedings of the Royal Society of London, Vol. A 110, pp. 561-579 (1926). Reprinted in B.L. van der Waerden. Sources of Quantum Mechanics.

3. On the Theory of Quantum Mechanics.

   Proc. Roy. Soc. London, Vol. A 112, pp. 661-677 (1926).

4. The Quantum Theory of the Emission and Absorption of Radiation.

Proc. Roy. Soc. London, Vol. A 114, pp. 243-265 (1927). Reprinted in J. Schwinger, *Quantum Electrodynamics*. New York: Dover, 1958.

5. The Quantum Theory of the Electron.

Proc. Roy. Soc. London, Vol. A 117, pp. 610-624 (1928).

Reprinted in C.W. Kilmister. *Special Theory of Relativity*. Oxford & New York: Pergamon Press, 1970.

6. The Basis of Statistical Quantum Mechanics.

Proc. Cambridge Phil. Soc. Vol. 25, pp. 62-66 (1929).

7. *The Principles of Quantum Mechanics*. Oxford: The Clarendon Press, 1930. 2nd ed. 1935. 3rd ed. 1947.

French trans. by A. Proca and J. Ullmo. *Les principes de la mécanique quantique*. Paris: Les Presses Universitaires de France, 1931.

Richard Lloyd DOAN (1898-)

See A.H. Compton.

Pierre DUHEM (1861-1916)

1. *Le potentiel thermodynamique et ses applications à la mécanique chimique et à la théorie des phénomènes électriques*. Paris: A. Hermann, 1886.

2. *Les Théories électriques de J. Clerk Maxwell: Étude historique et critique*. Paris: A. Hermann, 1902.

3. *La theorie physique: son objet, sa structure*. Paris: Marcel Rivière & Cie., 1906; 2nd ed. 1914.

English translation by Philip P. Wiener. *The Aim and Structure of Physical Theory*. Princeton University Press, 1954; paperback reprint by Atheneum, New York, 1962.

Freeman John DYSON (1923-)

1. The Radiation Theories of Tomonaga, Schwinger, and Feynman.

Phys. Rev., Series 2, Vol. 75, pp. 486-502 (1949).

Reprinted in J. Schwinger. *Quantum Electrodynamics*. New York: Dover Publications, 1958.

Paul EHRENFEST (1880-1933) and Tatiana EHRENFEST-AFANASSJEWA

1. Begriffliche Grundlagen der statistischen Auffassung in der Mechanik.

   Encyklopädie der mathematischen Wissenschaften, Bd. IV, Teil IV, Art. 32 (1911).

   English translation by Michael J. Moravcsik. The Conceptual Foundations of the Statistical Approach in Mechanics. Ithaca: Cornell University Press, 1959.

Paul EHRENFEST

2. Over adiabatische veranderingen van een stelsel in verband met de theorie der quanta.

   Verdagen Kon. Akad. Amsterdam, Vol. 25, pp. 412-433 (1916).

   English trans. "On Adiabatic Changes of a System in Connection with the Quantum Theory" in Proc. Sect. Sci. Acad. Amsterdam, Vol. 19, pp. 576-597 (1917).

   German trans. in Ann. Physik, Series 4, Vol. 51, pp. 327-352 (1916).

   Abridged English trans. "Adiabatic Invariants and the Theory of Quanta" in Phil. Mag., Series 6, Vol. 33, pp. 500-513 (1917); reprinted in B.L. van der Waerden, Sources of Quantum Mechanics. Amsterdam: North-Holland Pub. Co., 1957; New York: Dover Publications.

Albert EINSTEIN (1879-1955)

1. Über einen die Erzeugung und Verwandlung des Lichtes betreffenden heuristischen Gesichtspunkt.

   Ann. Physik, Ser. 4, Vol. 17, pp. 132-148 (1905).

   English translation in Henry A. Boorse & Lloyd Motz. The World of the Atom. New York: Basic Books, 1966.

   English translation (slightly different) in D. ter Haar. The Old Quantum Theory. Oxford: Pergamon Press, 1967.

   English translation by A.B. Arons and M.P. Peppard in American Journal of Physics, Vol. 33, pp. 367f. (1965)

2. Über die von der molekularkinetischen Theorie der Wärme geforderte Bewegung von in ruhenden Flüssigkeiten suspendierten Teilchen.

   Ann. Physik, Ser. 4, Vol. 17, pp. 549-560 (1905).

   Reprinted with notes by R. Fürth in Untersuchungen über die Theorie der Brownschen Bewegungen (Ostwalds Klassiker nr. 199). Leipzig: Akademische Verlagsgesellschaft, 1922.

English translation of the Fürth edition by A.D. Cowper. Investigations on the Theory of the Brownian Movement. London: Methuen, 1926; reprinted by Dover Publications, 1956.

3. Zur Elektrodynamik bewegter Körper.

   Ann. Physik, Ser. 4, Vol. 17, pp. 891-921 (1905).

   English translation "On the Electrodynamics of Moving Bodies" by W. Perrett and G.B. Jeffery in The Principle of Relativity . . . By H.A. Lorenz, A. Einstein, H. Minkowski, and H. Weyl. London: Methuen & Co., Ltd., 1923; Dover reprint. This translation is also reprinted in C.W. Kilmister. Special Theory of Relativity. Oxford & New York: Pergamon Press, 1970. See "Mistranslation of a passage in Einstein's original paper on relativity," by Charles Scribner Jr., in Amer.J.Phys.,Vol. 31, p. 398 (1963).

4. Ist die Trägheit eines Körpers von seinem Energieinhalt abhängig ?

   Ann. Physik, Ser. 4, Vol. 18, pp. 639-641 (1905).

   English translation, "Does the Inertia of a Body Depend on Its Energy-Content?" by W. Perrett and G.B. Jeffery in The Principle of Relativity . . . by H.A. Lorentz, A. Einstein, H. Minkowski and H. Weyl. London: Methuen, 1923; reprinted by Dover.

5. Über den Einfluss der Schwerkraft auf die Ausbreitung des Lichtes.

   Annalen der Physik, Ser. 4, Vol. 35, pp. 898-908 (1911).

   English translation, "On the Influence of Gravitation on the Propagation of Light" by W. Perrett and G.B. Jeffery, in The Principle of Relativity . . . by H.A. Lorentz, A. Einstein, H. Minkowski and H. Weyl. London: Methuen, 1923; reprinted by Dover.

6. Die Grundlage der allgemeinen Relativitätstheorie.

   Annalen der Physik, Ser. 4, Vol. 49, pp. 769-822 (1916).

   Also published separately under the same title with "Inhalt" and "Einleitung" added, by Barth. Leipzig, 1916.

   English translation by W. Perrett and G.B. Jeffery, in The Principle of Relativity . . . by H.A. Lorentz, A. Einstein, H. Minkowski and H. Weyl. London: Methuen, 1923; reprinted by Dover.

7. Über die spezielle und die allgemeine Relativitätstheorie, gemeinverstandlich.

   Vieweg: Braunschweig, 1917; 3rd ed., 1918; 10th ed., 1920; 14th ed., 1922.

   English translation by Robert W. Lawson from the 5th German edition. Relativity, The Special and The General Theory. London: Methuen, 1920; reprinted by Crown Publishers, New York, as: Relativity The Special and General Theory. 6th printing, 1960.

8. Zur quantentheorie der Strahlung.

   Physikalische Zeitschrift, Vol. 18, pp. 121-128 (1917).

   English translation "The Quantum Theory of Radiation" in Henry A. Boorse and Lloyd Motz. The World of the Atom. New York: Basic Books, 1966.

   English translation (slightly different) "On the Quantum Theory of Radiation" in D. ter Haar. The Old Quantum Theory. Oxford: Pergamon Press, 1967.

   Another English translation in B.L. van der Waerden. Sources of Quantum Mechanics. Amsterdam: North-Holland Pub. Co., 1967; New York: Dover Publications.

9. Vier Vorlesungen über Relativitätstheorie, gehalten im Mai 1921 an der Universität Princeton. Braunschweig: Vieweg, 1922.

   English translation by Edwin P. Adams. The Meaning of Relativity: Four Lectures Delivered at Princeton University, May, 1921. Princeton University Press, 1921; 2nd ed. with new appendices, The Meaning of Relativity. Princeton University Press, 1945.

   Russian translation by G.B. Itel'son. Matematichekija osnovy teori otnositel'nosti. Berlin: Slowo, 1923.

   French translation by Maurice Solovine. Quatre conférences sur la théorie de la relativité, faites à la Université de Princton. Paris: Gauthier, 1924.

10. Quantentheorie des einatomigen idealen Gases.

    Sitzungsber. Akad. Wiss., Berlin, 1924, pp. 261-267. ". . . Zweite Abhandlung," ibid., 1925, pp. 3-25.

    English translation by J. Mehra. "Quantum Theory of the Monatomic Ideal Gas" (to be published).

Albert EINSTEIN, B. PODOLSKY, and S. ROSEN

11. Can Quantum-Mechanical Description of Physical Reality Be Considered Complete?

Physical Review, Ser. 2, Vol. 47, pp. 777-780 (1935).

Reprinted in *Physical Reality*, ed. S. Toulmin. New York: Harper & Row, 1970. With reply by N. Bohr. First published under the same title in Physical Review, Ser. 2, Vol. 48, pp. 696-702 (1935).

David ENSKOG (1884-1947)

1. *Kinetische Theorie der Vorgänge in mässig verdünnten Gasen*.

   Uppsala: Almqvist & Wiksell Boktryckeri AB, 1917.

   English trans. by J. Kopp in S.G. Brush. *Kinetic Theory*. Vol. 3. Oxford: Pergamon Press, 1972.

2. Kinetische Theorie der Wärmeleitung, Reibung und Selbstdiffusion in gewissen verdichteten Gasen und Flüssigkeiten.

   Kungliga Svenska Vetenskapsakademiens Handlingar, Ny Foljd, Vol. 63, No. 4 (1922).

   English translation by J. Kopp in S.G. Brush. *Kinetic Theory*. Vol. 3.

Leonhard EULER (1707-1783)

1. Dissertatio physica de sono.

   Basel, 1727.

   Reprinted in his *Opera Omnia*.

2. De motu vibratorio typanorum.

   Acad. Sci. St. Petersburg, Vol. 10, pp. 243- (1746).

3. De la propagation du son.

   Mem. Acad. Sci. Berlin, Vol. 15, pp. 185-209 (1759).

   Reprinted in *Leonhardi Euleri Opera Omnia* (III). Leipzig: B.G. Teubner, 1926. Vol. I.

4. Éclaircissements plus détailles sur la generation et la propagation du son, et sur la formation de l'écho.

   Mem. Acad. Sci. Berlin, Vol. 21, pp. 335-364 (1765).

   Reprinted in his *Opera Omnia* (III). Vol. 1.

5. *Lettres á une princesse d'Allemagne sur divers sujets de physique et de philosophie*.

   St. Petersburg, 1768-1772 (3 vols.).

English trans., Letters of Euler on Different Subjects in Natural Philosophy Addressed to a German Princess. London, 1802.

German trans., Briefe an eine deutsche Prinzessin über verschiedene Gegenstände aus der Physik und Philosophie. Leipzig, 1769, 1773.

Michael FARADAY (1791-1867)

1. Experimental Researches in Electricity. 3 vols.

   Vols. I and III first pub. 1839 and 1845 by Taylor and Francis. Vol. II first pub. 1844 by R. and J.E. Taylor.

   Reprinted by Dover Pubs., N.Y., 1965 (3 vols. in 2, total $15.00).

   German translations in Annalen der Physik, Series 2, Vols. 25-69. Reprinted in Experimental Untersuchungen über Elektricität, hrsg. A.J. von Oettingen. Leipzig: W. Engelmann, 1896-1903; Ostwalds Klassiker nr. 81, 86, 87, 126, 128, 131, 134, 136, 140.

Berend Wilhelm FEDDERSEN (1832-1918)

1. Beiträge zur Kenntnis des elektrischen Funkens.

   Annalen der Physik, Vol. 179 [= Ser. 2, Vol. 103], pp. 69-88 (1858).

Enrico FERMI (1901-1954)

1. Sulla quantizzazione del gas perfetto monatomico.

   Rendiconti della Reale Accademia dei Lincei, Vol. 3, pp. 145-149 (1926). Presented at the meeting of 7 February 1926.

   Reprinted in The Collected Papers of Enrico Fermi. Vol. 1. Chicago & London: The University of Chicago Press, and Rome: Accademia Nationale dei Lincei, 1962.

   This is a brief announcement of the theory developed in greater detail in Fermi's paper in the Zeitschrift für Physik (q.v.).

2. Zur Quantelung des idealen einatomigen Gases.

   Z.f. Physik, Vol. 36, pp. 902-912 (1926).

   Reprinted in The Collected Papers of Enrico Fermi. Vol. 1. Chicago & London: The University of Chicago Press, and Rome: Accademia Nazionale dei Lincei, 1962.

This is an extended version of the earlier announcement published in Italian.

English translation (omitting footnotes and last paragraph) in H.A. Boorse & Lloyd Motz. The World of the Atom. New York: Basic Books, 1966.

3. Versuch einer Theorie der β-Strahlen.

   Z.f. Physik, Vol. 88, pp. 161-177 (1934).

   This appears to be Fermi's own German version of a paper published about the same time in Italian: "Tentativo di una teoria dei ragg β."

   Nuovo Cimento, Vol. 11, pp. 1-19 (1934).

   Both are reprinted in The Collected Papers of Enrico Fermi. Vol. 1. Chicago & London: The University of Chicago Press, and Rome: Accademia Nazionale dei Lincei, 1962.

   German version reprinted in R.T. Beyer, Foundations of Nuclear Physics. New York: Dover Publications, 1949.

   English trans. (from the German version), "Attempt at a Theory of Beta Rays" by Fred L. Wilson, in Amer. J. Phys., Vol. 36, pp. 1150-1160 (1968).

   Another English trans., "Tentative Theory of Beta-Radiation," in Charles Strachan. The Theory of Beta-Decay. Oxford: Pergamon Press, 1969.

Sidney FERNBACH (1917-), R. SERBER, and T.B. TAYLOR

1. The Scattering of High-Energy Neutrons by Nuclei.

   Physical Review, Series 2, Vol. 75, pp. 1352-1355 (1949).

   Reprinted in I.E. McCarthy. Nuclear Reactions. Oxford: Pergamon Press, 1970.

Herman FESHBACH (1917-), C.E. PORTER and V.F. WEISSKOPF

1. Model for Nuclear Reactions with Neutrons.

   Physical Review, Series 2, Vol. 96, pp. 448-464 (1954).

   Reprinted in I.E. McCarthy. Nuclear Reactions. Oxford: Pergamon Press, 1970.

Richard Phillips FEYNMAN (1918-)

1. Space-Time Approach to Non-Relativistic Quantum Mechanics.

   Reviews of Modern Physics, Vol. 20, pp. 367-387 (1948).

Reprinted in J. Schwinger. Quantum Electrodynamics. New York: Dover Publications, 1958.

2. The Theory of Positrons.

   Phys. Rev., Series 2, Vol. 76, pp. 749-759 (1949).

   Reprinted in J. Schwinger. Quantum Electrodynamics. New York: Dover Publications, 1958.

3. Space-Time Approach to Quantum Electrodynamics.

   Phys. Rev., Series 2, Vol. 76, pp. 769-789 (1949).

   Reprinted in J. Schwinger. Quantum Electrodynamics. New York: Dover Publications, 1958.

4. Mathematical Formulation of the Quantum Theory of Electromagnetic Interaction.

   Phys. Rev., Series 2, Vol. 80, pp. 440-457 (1950).

   Reprinted in J. Schwinger. Quantum Electrodynamics. New York: Dover Publications, 1958.

R.P.FEYNMAN and M. GELL-MANN

5. Theory of the Fermi Interaction.

   Phys. Rev., Ser. 2, Vol. 109, pp. 193-198 (1958).

   Reprinted in Charles Strachan. The Theory of Beta Decay. Oxford: Pergamon Press, 1969.

George Francis FITZGERALD (1851-1901)

1. The Ether and the Earth's Atmosphere.

   Science, Vol. 13, p. 390 (1889).

   Reprinted in S.G. Brush. "Note on the History of the Fitzgerald-Lorentz Contraction," Isis, Vol. 58, pp. 230-232 (1967).

Armand Hippolyte Louis FIZEAU (1819-1896) and L. BREGUET

1. Sur l'expérience relative à la vitesse comparative de la lumiére dans l'air et dans l'eau.

   Compt. Rend. Acad. Sci. Paris, Vol. 30, pp. 771-774 (1850).

Jean Bernard Léon FOUCAULT (1819-1868)

1. Méthode génerale pour mesurer la vitesse de la lumière dans l'air et les milieux transparents. Vitesses

relatives de la lumière dans l'air et dans l'eau. Projet d'experience sur la vitesse de propagation du calorique rayonnant.

Compt. Rend. Acad. Sci. Paris, Vol. 30, pp. 551-560 (1850).

German trans. in Annalen der Physik, Vol. 157 [= Ser. 2, Vol. 81], pp. 434-442 (1850).

Jean Baptiste Joseph FOURIER (1768-1830)

1. Mémoire su la propagation de la chaleur dans les corps solides.

   Bull. Société Philomathique (Paris), Vol. 1, pp. 112-116 (1808).

   Reprinted in Fourier's Oeuvres. Ed., Gaston Darboux. Paris: Gauthier-Villars, 1888 & 1890.

2. Théorie du mouvement de la chaleur dans les corps solides. Submitted 28 September 1811 for prize of the Academie des Sciences.

   First published in Mémoires de l'Academie Royale des Sciences. Paris, Vol. 4, pp. 185-555 (1824).

   A second part, "Suite du mémoire intitule: Théorie du mouvement de la chaleur dans les corps solides," was published in Mem. Acad. Roy. Sci (for 1821 and 1822), Vol. 5, pp. 153-246 (1826).

   Revised version of the first part published separately in 1822 under the title, Théorie Analytique de la Chaleur. Paris: Chez Firmim Didot.

   This and the second part of the original memoir are reprinted in the Oeuvres de Fourier. Ed., Darboux. Paris, 1888, 1890.

   The first part only was translated into English by Alexander Freeman. The Analytical Theory of Heat. London, 1878; reprinted by Dover Publications, New York, 1955.

James FRANCK (1882-1964) and Gustav HERTZ (1887-)

1. Über Zusammenstösse zwischen Elektronen und den Molekülen des Quecksilberdampfes und die Ionisierungsspannung desselben.

   Verhand. Deut. Physik. Ges. Vol. 16, pp. 457-467 (1914).

   English translation, "Collisions between Electrons and Mercury Vapor Molecules and the Ionization Potential of such Molecules," in Henry A. Boorse and Lloyd Motz, The World of the Atom. New York: Basic Books, 1966.

2. Über die Erregung der Quecksilberresonanzlinie 253,6 μμ durch Elektronenstösse.

   Verhand. Deut. Physik. Ges. Vol. 16, pp. 512-517 (1914).

   English translation, "On the Excitation of the 2536 Å Mercury Resonance Line by Electron Collisions," in D. ter. Haar. The Old Quantum Theory. Oxford: Pergamon Press, 1967.

Benjamin FRANKLIN (1706-1790)

1. Experiments and Observations on Electricity, Made at Philadelphia in America. London, 1751-1753 (2 vols.).

   Reprint, edited by I.B. Cohen. Harvard University Press, 1941.

   French trans., by M. d'Alibard. Experiences et observations sur l'électricité. Paris, 1756.

Augustin FRESNEL (1788-1827)

1. Sur l'influence du mouvement terrestre dans quelques phénomènes d'optique.

   Ann. Chim., Vol. 9, pp. 57-66, 286 (1818).

   Reprinted in his Oeuvres Complètes. Paris, 1866-1870. Vol. 2.

2. Mémoire sur la diffraction de la lumière.

   Mem. Acad. Sci. Paris, Vol. 5, pp. 339-475 (1826).

   Reprinted in his Oeuvres. Vol. 1.

   German trans. in Ann. Physik, Series 2, Vol. 30, pp. 100-261 (1836).

   English trans. (partial) in H. Crew. The Wave Theory of Light. New York: American Book Co., 1900.

Walter FRIEDRICH (1883-), P. KNIPPING & M. LAUE

1. "Interferenz-Erscheinungen bei Röntgenstrahlen."

   Bayerische Akademie der Wissenschaften zu München, Sitzungsberichte math.-phys. Klasse (1912), pp. 303-322.

   English translation, "X-Ray Interference Phenomena" in C.F. Bacon. X-Ray and Neutron Diffraction. Oxford: Pergamon Press, 1966.

Galileo GALILEI (1564-1642)

1. De Motu (ca. 1590).

   English translation by I.E. Drabkin "On Motion" in On Motion and On Mechanics. Madison: University of Wisconsin Press, 1960.

2. Le Meccaniche (ca. 1600).

   English translation by S. Drake in "On Mechanics" in On Motion and On Mechanics. Madison: University of Wisconsin Press, Madison, 1960.

3. Discorsi e dimostrazioni matematichè intorno à due nuoue scienze, attenenti alla Mecanica & i Movimenti Locali. Leiden 1638.

   English translation by Henry Crew and Alfonso De Salvio. Dialogues Concerning Two New Sciences. New York: Macmillan, 1914; reprinted by Dover Publications.

   German trans. ed. by Arthur von Oettingen. Unterredungen und mathematische Demonstrationen über zwei neue Wissenszweige, die Mechanik und die Fallgesetze betreffend. Leipzig: W. Engelman, 1890-1891; Ostwalds Klassiker nr. 11, 24, 25.

George GAMOW (1904-1968)

1. Zur Quantentheorie des Atomkerne.

   Z.f. Physik, Vol. 51, pp. 204-212 (1928).

   Reprinted in R.T. Beyer. Foundations of Nuclear Physics. New York: Dover Publications, Inc., 1949.

G. GAMOW and E. TELLER

2. Selection Rules for the Beta-Disintegration.

   Phys. Rev., Series 2, Vol. 49, pp. 895-899 (1936).

   Reprinted in Charles Strachan. The Theory of Beta-Decay. Oxford: Pergamon Press, 1969.

Hans GEIGER (1882-1945) and E. MARSDEN

1. On a Diffuse Reflection of the Alpha Particles.

   Proc. Roy. Soc. London, Vol. A82, pp. 495-500 (1909).

   Reprinted in Stephen Wright. Classical Scientific Papers —Physics. New York: American Elsevier Pub. Co., 1965.

2. The Laws of Deflexion of Alpha Particles through Large Angles.

   Phil. Mag., Series 6, Vol. 27, pp. 604-623 (1913).

Murray GELL-MANN (1929-)

See R.P. FEYNMAN.

Walter GERLACH (1889-) and Otto STERN

1. Der experimentelle Nachweis der Richtungsquantelung im Magnetfeld.

   Z.f. Physik, Vol. 9, pp. 349-352 (1922).

   English translation "Experimental Proof of Space Quantization in a Magnetic Field" in Henry A. Boorse & Lloyd Motz. The World of the Atom. New York: Basic Books, 1966.

Lester Halbert GERMER (1896-1971)

See C.J. DAVISSON.

Josiah Willard GIBBS (1839-1903)

1. On the Equilibrium of Heterogeneous Substances.

   Trans. Conn. Acad., Vol. 3, pp. 108-248 (1876) and pp. 343-524 (1878).

   Reprinted in The Collected Works of J. Willard Gibbs. 1906; 2nd ed. 1928; reprinted by Dover Publications, New York, 1960.

   German trans. by W. Ostwald. Thermodynamische Studien. Leipzig: Engelmann, 1892.

   French trans. by Georges Matisse. L'équilibre des substances hétérogènes. Paris: Gauthier-Villars, 1919.

2. Elementary Principles in Statistical Mechanics, Developed with Especial Reference to the Rational Foundation of Thermodynamics. New York: Scribner, 1902.

   Reprinted in The Collected Works of J. Willard Gibbs. London & New York: 2nd ed., 1928; reprinted by Dover Publications, New York, 1960.

   German translation ed. by E. Zermelo: Elementare Grundlagen der statistischen Mechanik. Leipzig: Barth, 1905.

French translation by F. Cosserat; rev. & completed by J. Rossignol. Principes élémentaires de mécanique statistique . . . Paris: J. Hermann, 1926.

Russian translation by K.V. Nikolski. Osnovanye Printsipy Statisticheskoi Mekhaniki . . . Moscow: Gos. Izd.-vo Tekhniko-Teoreticheskoi Literatury, 1946.

William GILBERT (1540-1603)

1. De Magnete, Magneticisque corporibus, et de magno magnete tellure; Physiologia nova plurimus & argumentis, & experimentis demonstrata. London, 1600.

   English translation by P. Fleury Mottelay. On the Loadstone and Magnetic Bodies, and on the Great Magnet the Earth. A New Physiology, Demonstrated with Many Arguments and Experiments. New York: J. Wiley & Sons, 1893; reprinted by Dover Publications, 1958.

   Russian trans. by A.I. Dobatur, ed. A.G. Kalashnikov, O Magnite . . . Moscow: Izd.-vo Akademii Nauk SSSR, 1956.

2. De Mundo nostro sublunari philosophia nova. Amsterdam: Elzevier, 1651.

   Reprinted by Menno Hertzberger, Amsterdam, 1965; with Sister Suzanne Kelly, The De Mundo of William Gilbert.

Vitalii Lazarevich GINZBURG (1916-) and L.D. LANDAU

1. K teorii sverkhlrovodimosti.

   Zhur. eksp. teor. fiz., Vol. 20, pp. 1064-1082 (1950).

   English translation "On the Theory of Superconductivity" in D. ter Haar (ed.). Collected Papers of L.D. Landau. Oxford: Pergamon Press, 1965.

Johann Wolfgang von GOETHE (1749-1832)

1. Zur Farbenlehre. Tübingen, 1810.

   English translation by C.L. Easlake. Theory of Colours. Part 1 only, omitting most of the polemical material attacking Newton's theory. London: John Murray, 1840; reprinted by MIT Press, Cambridge, 1970.

S. GOUDSMIT

See G.E. UHLENBECK.

Francesco Maria GRIMALDI (1613-1663)

1. Physico-Mathesis de lumine, coloribus, et iride. Bologna, 1665.

Otto von GUERICKE (1602-1686)

1. Experimenta nova (ut vocantur) magdeburgica de vacuo spatio primum à R.P. Gaspare Schotto . . . nunc verò ab ipso auctore perfectiùs edita, variisque aliis experimentis aucta. Quibus accesserunt simul certa qaedam de aëris pondere circa terram; de virtutibus mundanis, & systemate mundi planetario; sicut & de stellis fixis, ac spatio illo immenso, quod tam intra quam extra eas funditur. Amsterdam, 1672.

   German translation by F. Dannemann. Otto von Guericke's neue Magdeburgische Versuche über den leeren Raum. Ostwald's Klassiker, nr. 59. Leipzig: W. Engelmann, 1894.

   German translation by Hans Schimank. Otto von Guerickes Neue (sogenannte) Magdeburger Versuche über den leeren Raum. Nebst Briefen, Urkunden und anderen Zeugnissen. Düsseldorf: VDI-Verlag, 1968.

William Rowan HAMILTON (1805-1865)

1. On a general method in dynamics, by which the study of the motions of all free systems of attracting or repelling points is reduced to the search and differentiation of one central function, or characteristic function.

   Phil. Trans. Roy. Soc. London, Part II for 1834, pp. 247-308.

   French trans. in Correspondance Mathematique et Physique, Vol. 8, pp. 69-89, 200-211 (1834).

2. Second essay on a general method in dynamics.

   Phil. Trans. Roy. Soc. London, Part I for 1835, pp. 95-144.

R.W. HAYWARD

See C.S. WU

Werner HEISENBERG (1901-)

1. Über quantentheoretische Umdeutung kinematischer und mechanischer Beziehungen.

   Z.f. Physik, Vol. 33, pp. 879-893 (1925).

English translation, "The Interpretation of Kinematic and Mechanical Relationships According to the Quantum Theory" in G. Ludwig. Wave Mechanics. Oxford: Pergamon Press, 1968.

Another English trans. "Quantum-Theoretical Re-Interpretation of Kinematic and Mechanical Relations" in B.L. van der Waerden. Sources of Quantum Mechanics. Amsterdam: North-Holland Pub. Co., 1967; New York: Dover Publications.

2. Die physikalischen Prinzipien der Quantentheorie. Leipzig: S. Hirzel, 1930.

   English translation by Carl Eckart and Frank C. Hott. The Physical Principles of the Quantum Theory. University of Chicago, 1930; Dover reprint, 1949?

   French trans. by B. Champion and E. Hochard. Les Principes Physiques de la Théorie des Quanta. Paris: Gauthier-Villars, 1957.

   Italian trans. by Mario Agano. I Principi fisici della teoria dei quanti. Torino: G. Einaudi, 1948.

3. Über den Bau der Atomkerne I, II, III.

   Zeits. f. Physik, Vol. 77, pp. 1-11 (1932); Vol. 78, pp. 156-164 (1932); Vol. 80, pp. 587-596 (1939).

   English translation "On the Structure of Atomic Nuclei." Parts I and III in D.M. Brink. Nuclear Forces. Oxford: Pergamon Press, 1965.

Hermann von HELMHOLTZ (1821-1894)

1. Ueber die Erhaltung der Kraft,Eine physikalische Abhandlung. Berlin: G. Reimer, 1847.

   English translation by John Tyndall in Taylor's Scientific Memoirs (1853).

   Tyndall's translation reprinted in part in S.G. Brush. Kinetic Theory, Vol. 1. Oxford: Pergamon Press, 1965.

John HERAPATH (1790-1868)

1. A mathematical inquiry into the causes, laws, and principal phenomena of heat, gases, gravitation, etc.

   Annals of Philosophy, Series 2, Vol. 1, pp. 273-293, 340-351, 401-416 (1821). Reprinted in Mathematical Physics and Selected Papers by John Herapath. New York: Johnson Reprint Corp., 1972.

2. Mathematical Physics. London, 1847. Reprinted in Mathematical Physics and Selected Papers . . .

Heinrich HERTZ (1857-1894)

1. Über Strahlen elektrischer Kraft.

   Sitzungsber. Preuss. Akad. Wiss. Berlin, Dec. 1888, pp. 1297-1307.

   Reprinted with other papers in Untersuchungen über die Ausbreitung der Elektrische Kraft (1892).

   English translation by D.E. Jones in Electric Waves. London: Macmillan, 1893; reprinted by Dover Pubs., Inc., 1962.

David HILBERT (1862-1943)

1. Begründung der kinetischen Gastheorie.

   Chapter XXII in his Grundzüge einer allgemeinen Theorie der linearen Integralgleichungen. Leipzig & Berlin: Teubner, 1912.

   Reprinted in Mathematische Annalen, Vol. 72, pp. 562-577 (1912).

   English translation by J. Kopp in S.G. Brush. Kinetic Theory. Vol. 3. The Chapman-Enskog solution of the transport equation for moderately dense gases. Oxford: Pergamon Press, 1972.

Robert HOOKE (1635-1703)

1. Lectures de Potentia Restitutiva, or of Spring. Explaining the power of springing bodies. London, 1678.

D.D. HOPPES

See C.S. WU.

R.P. HUDSON

See C.S. WU.

Christiaan HUYGENS (1629-1695)

1. Sur le mouvement, qui est produit par la rencontre des corps.

   Journal des Savants, 1669.

   "The laws of motion on the collision of bodies." Translated from the Latin Philosophical Transactions, No. 46, p. 925 (1669).

p. 925 (1669). In I.B. Cohen. *Readings in the Physical Sciences*. Cambridge, Mass., 1966.

2. *Horologium Oscillatorium*. Paris, 1673.

   Reprinted in his *Oeuvres Completes*, t. 18, with French translation on facing pages.

   German trans. by A. Heckscher and A.V. Oettingen. *Die Pendeluhr*. Leipzig: W. Engelmann, 1913; Ostwalds *Klassiker* nr. 192.

3. *Traité de la Lumière*. Communicated to the Academie des Sciences, Paris, 1678. Published with revisions in 1690.

   English translation by Silvanus P. Thompson. *Treatise on Light*. London: Macmillan, 1912; reprinted by Dover Publications, 1962.

   German trans. by E. Lommel. *Abhandlung über das Licht*. Leipzig: W. Engelmann, 1890; Ostwalds *Klassiker* nr. 20.

4. De Motu Corporum percussione (published posthumously in 1703), in his *Opera Reliqua*. Amsterdam, 1728; reprinted in *Oeuvres Complètes*. Vol. 16. The Hague, 1888-1950.

   English translation "On the Movement of Bodies through Impact" by J.E. Murdoch, in I.B. Cohen. *Readings in the Physical Sciences*. Cambridge, Mass., 1966.

   German trans. "Über die Bewegung der Körper durch den stoss." Ed., Felix Hausdorff. In *Christiaan Huygens nachgelassene Abhandlungen*. Leipzig: W. Engelmann, 1903; Ostwalds *Klassiker* nr. 138.

5. *Theoremata de vi centrifuga et motu circulari demonstrata*, appended to J. Keill's *Introductio ad veram physicam*, etc. (1719), pp. 255-274.

   English translation in the 1745 edition of this book.

Ernst ISING (1900-)

1. Beitrag zur Theorie des Ferromagnetismus.

   Zeits. f. Physik, Vol. 31, pp. 253-258 (1925).

   Reprinted in K.B. Beaton & H.C. Bolton. *A German Source-Book in Physics*. Oxford: Clarendon Press, 1969.

H.A. JAHN and Edward TELLER

1. Stability of Polyatomic Molecules in Degenerate Electronic States. I. Orbital Degeneracy.

Proceedings of the Royal Society of London. Vol. A 161, pp. 220-235 (1937).

Reprinted in Arthur P. Cracknell. Applied Group Theory. Oxford: Pergamon Press, 1968.

Frédêric JOLIOT-CURIE (1900-1958)

See Irene JOLIOT-CURIE.

Irene JOLIOT-CURIE (1897-1956) and Frédêric JOLIOT-CURIE

1. "Un nouveau type de radioactivitê."

   Compt. Rend. Acad. Sci. Paris, Vol. 198, pp. 254-256 (1934).

   Reprinted in R.T. Beyer. Foundations of Nuclear Physics. New York: Dover Publications, Inc., 1949.

[Ernest] Pascual [Wilhelm] JORDAN (1902-)

See M. BORN.

James Prescott JOULE (1818-1889)

1. On the calorific effects of magneto-electricity, and on the mechanical value of heat.

   Phil. Mag., Ser. 3, Vol. 23, pp. 263-276, 347-355, 435-443 (1843).

   Reprinted in The Scientific Papers of James Prescott Joule. London, 1884.

2. On the existence of an equivalent relation between heat and the ordinary forms of mechanical power.

   Philosophical Magazine, Series 3, Vol. 27, pp. 205-207 (1845).

   Reprinted in The Scientific Papers of James Prescott Joule. London, 1884.

3. On the mechanical equivalent of heat as determined by the heat evolved by the friction of fluids.

   Phil. Mag., Ser. 3, Vol. 31, pp. 173-176 (1847).

   German trans. in Annalen der Physik, Vol. 149 [= Ser. 2, Vol. 73], pp. 479-484 (1848).

4. On matter, living force, and heat.

   Manchester Courier, 5 and 12 May, 1847; reprinted in The Scientific Papers of James Prescott Joule. London: The Physical Society, 1884.

   Reprinted in S.G. Brush. Kinetic Theory. Vol. 1. Oxford: Pergamon Press, 1965.

5. Some remarks on heat, and the constitution of elastic fluids.

   Memoirs of the Lit. and Phil. Soc. of Manchester, Ser. 2, Vol. 9, pp. 107-114 (1851, read 1848).

   Reprinted with a note, in Phil. Mag., Ser. 4, Vol. 14, pp. 211- (1857).

   Reprinted in The Scientific Papers of James Prescott Joule. London, 1884.

   (Partial) French trans. in Ann. Chim., Vol. 50, pp. 381-383 (1857).

Heike KAMERLINGH ONNES (1853-1926) and J.D.A. BOKS

1. Further Experiments with Liquid Helium. V. The Variation of Density of Liquid Helium Below the Boiling Point.

   Communications of the Physical Laboratory. Leiden, nr. 170b (1924).

   Reprinted in Z.M. Galasiewicz. Helium 4. Oxford: Pergamon Press, 1971.

Wilhelmus Hendrikus KEESOM (1876- ?) and M. WOLFKE

1. Two Different States of Helium.

   Communications of the Physical Laboratory. Leiden, nr. 190b (1927).

   Reprinted in Z.M. Galasiewicz. Helium 4. Oxford: Pergamon Press, 1971.

Wilhelmus Hendrikus KEESOM and K. CLUSIUS

2. Specific Heat of Liquid Helium.

   Communications of the Physical Laboratory. Leiden, nr. 219e.

   Reprinted in Z.M. Galasiewicz. Helium 4. Oxford: Pergamon Press, 1971.

William Thomson, Lord KELVIN (1824-1907)

1. On an absolute thermometric scale founded on Carnot's theory of the motive power of heat, and calculated from Regnault's observations.

   Phil. Mag., Series 3, Vol. 33, pp. 313- (1848).

   Proc. Cambridge Phil. Soc., Vol. 1, pp. 66-71 (1848; pub. 1866).

   Reprinted in his Mathematical and Physical Papers. Vol. I. Cambridge, 1882.

2. An account of Carnot's theory of the motive power of heat; with numerical results deduced from Regnault's experiments on steam.

   Trans. Roy. Soc. Edinburgh, Vol. 16, pp. 541-574 (1849).

   Reprinted in his Mathematical and Physical Papers. Vol. I.

3. On a universal tendency in nature to the dissipation of energy.

   Phil. Mag., Series 4, Vol. 4, pp. 304-306 (1852).

4. On the dynamical theory of heat, with numerical results deduced from Mr. Joule's equivalent of a thermal unit, and M. Regnault's observations on steam.

   Trans. Roy. Soc. Edinburgh, Vol. 20, pp. 261-288 (1853).

   Phil. Mag., Series 4, Vol. 4, pp. 8-21, 105-117, 168-176, 424-434 (1852); Vol. 9, pp. 523-531 (1855); Vol. 11, pp. 214-225, 281-297, 379-388, 433-446 (1856).

   Reprinted in his Mathematical and Physical Papers, Vol. I, pp. 174-210. Cambridge, 1882.

5. The size of atoms.

   Nature, Vol. 1, p. 551 (1870).

   Reprinted in Thomson and Tait's Treatise on Natural Philosophy; reprinted as Principles of Mechanics and Dynamics by Dover Publications.

   French trans. in Les Mondes, Vol. 22, pp. 701-708 (1870); see also Vol. 27, pp. 616-623 (1872).

   German trans. in Ann. Chem. Pharm., Vol. 157, pp. 54-66 (1871).

6. Nineteenth century clouds over the dynamical theory of heat and light. (Friday evening lecture, Royal Institution, April 27, 1900.)

Proc. Roy. Inst., Vol. 16, pp. 363-397 (1902).

Phil. Mag., Ser. 6, Vol. 2, pp. 1-40 (1901).

Reprinted with additions, as Appendix B, to his _Baltimore Lectures on Molecular Dynamics and the Wave Theory of Light_. London: Clay, 1904 (pp. 486-527).

Johannes KEPLER (1571-1630)

1. _Ad Vitellionem Paralipomena, quibus Astronomiae pars optica traditur; potissimum de artificiosa observatione et aestimatione diametrorum deliquiorumque solis et lunae. Cum exemplis insignium eclipsium_. Frankfurt, 1604.

   Reprinted in Kepler's _Opera Omnia_. Ed., C. Frisch. Vol. II, pp. 119-397.

   Also in Johannes Kepler _Gesammelte Werke_, Band II (hrsg. F. Hammer). München: C.H. Beck'sche Verlagsbuchlandlung, 1939.

   German translation by Ferdinand Plehn. Ed., Moritz von Rohr. _J. Kepler's Grundlagen der geometrischen Optik (im Anschluss an die Optik des Witelo)_. Leipzig: Akademische Verlagsgesellschaft m.b.H., 1922; Ostwald's _Klassiker_ nr. 198.

Aleksandr Iakovlevish KHINCHIN (1894-)

1. _Matematicheskie Osnovaniya Statisticheskoi Mekhaniki_. Moscow: Gos. Izd-vo Tekhn-Teoretich. Lit-ry, 1943.

   English translation by G. Gamow. _Mathematical Foundations of Statistical Mechanics_. New York: Dover Publications, 1949.

Gustav Robert KIRCHHOFF (1824-1887)

1. Über den Zusammenhang zwischen Emission und Absorption von Licht und Wärme.

   Monatsberichte der Akademie der Wissenschaften zu Berlin 1860, pp. 783-787.

   Reprinted in Kirchhoff's _Gesammelte Abhandlungen_. Leipzig, 1882; reprinted in Ostwalds _Klassiker_ nr. 100. Leipzig: Engelmann, 1898.

   English trans. in D.B. Brace. _The Laws of Radiation and Absorption_. New York: American Book Co., 1901.

2. Ueber einem neuen Satz der Wärmelehre.

   Verhandl. Nat. Med. Ver., pp. 16-23 (1859-1860).

   Reprinted in Dinglers Polytechn. J., Vol. 157, pp. 29-36 (1860).

English trans. "On a new proposition in the theory of heat" in Phil. Mag., Series 4, Vol. 21, pp. 241-247 (1861), reprinted in A.J. Meadows. Early Solar Physics. Oxford: Pergamon Press, 1970.

Paul KNIPPING (1883-)

See W. FRIEDRICH

Karl Rudolph KOENIG (1832-1901)

1. Quelques expériences d'acoustique. Paris: Lahure, 1882.

Hendrik Anthony KRAMERS (1894-1952)

See N. BOHR.

Joseph Louis LAGRANGE (1736-1813)

1. Recherches sur la nature et la propagation du son.

   Reprinted in Oeuvres de Lagrange. Paris, 1867. Vol. 1.

2. Mécanique analytique. Paris, 1788; new ed., 1811-1815.

   Reprint, combining notes from the 3rd revision, corrected and annotated by Joseph Bertrand, and the 4th ed. pub. under the direction of C. Darboux. Paris: Albert Blanchard, 1965. 2 vols.

Willis Eugene LAMB, Jr. (1913-) and Robert C. RETHERFORD

1. Fine Structure of the Hydrogen Atom by a Microwave Method.

   Phys. Rev., Series 2, Vol. 72, pp. 241-243 (1947).

   Reprinted in J. Schwinger. Quantum Electrodynamics. New York: Dover Publications, 1958.

Lev Davidovich LANDAU (1908-1968)

1. Diamagnetismus der Metalle.

   Zeits. f. Physik, Vol. 64, pp. 629-637 (1930).

   English translation in D. ter Haar (ed.). Collected Papers of L.D. Landau. Oxford: Pergamon Press, 1965.

2. Teoriya sverkhtekuchesti geliya II.

   Zhur. eksp. teor. fiz., Vol. 11, pp. 592- (1941).

English trans. "The Theory of Superfluidity of Helium II" in D. ter Haar (ed.). Collected Papers of L.D. Landau. Oxford: Pergamon Press, 1965.

3. Teoriya fermi-shidkosti.

   Zhur. eksp. teor. fiz., Vol. 30, pp. 1058-1064 (1956).

   English trans. "The Theory of a Fermi Liquid" in Soviet Physics JETP, Vol. 3, pp. 920- (1957), reprinted in D. ter Haar (ed.). Collected Papers of L.D. Landau.

See also V.L. GINZBURG

Alfred LANDÉ (1888-)

1. Termstruktur und Zeemaneffekt der Multiplette.

   Zeits. f. Physik, Vol. 15, pp. 189-205 (1923).

   English translation by J.B. Sykes. "Term Structure and Zeeman Effect in Multiplets." In W.R. Hindmarsh, Atomic Spectra. Oxford: Pergamon Press, 1967.

Paul LANGEVIN (1872-1946)

1. Sur la theorie du mouvement brownien.

   Compt. rend. Acad. Sci. Paris, Vol. 146, pp. 530-533 (1908).

   Reprinted in Oeuvres Scientifiques de Paul Langevin. Paris: Centre National de la Recherche Scientifique, 1950.

Pierre-Simon de LAPLACE (1749-1827)

See A. - L. LAVOISIER.

Max Theodor Felix von LAUE (1879-1960)

See W. FRIEDRICH.

Antoine-Laurent LAVOISIER (1743-1794) and Pierre-Simon de LAPLACE

1. Memoire sur la chaleur.

   Mem. Acad. Sci. Paris, 1780, pp. 355-.

   Reprinted in the series Les Maitres de la Pensée Scientifique. Paris: Gauthier-Villars, 1920.

Ernest Orlando LAWRENCE (1901-1958) and M.S. LIVINGSTON

1. "The Production of High Speed Light Ions Without the Use of High Voltages."

   Physical Review, Series 2, Vol. 40, pp. 19-35 (1932).

   Reprinted in R.T. Beyer. Foundations of Nuclear Physics. New York: Dover Publications, Inc., 1949.

Tsung-Dao LEE (1926-) and C.N. YANG

1. Question of Parity Conservation in Weak Interactions.

   Physical Review, Series 2, Vol. 104, pp. 254-258 (1956).

   Reprinted in Charles Strachan. The Theory of Beta-Decay. Oxford: Pergamon Press, 1969.

2. Parity Nonconservation and a Two-Component Theory of the Neutrino.

   Physical Review, Series 2, Vol. 105, pp. 1671-1675 (1957).

   Reprinted in Charles Strachan. The Theory of Beta-Decay. Oxford: Pergamon Press, 1969.

Gottfried Wilhelm LEIBNIZ (1646-1716)

1. Brevis demonstratio erroris memorabilis Cartesii et aliorum circa legem naturae, secundum quam volunt a Deo eandem semper quantitatem motus conservari, qua et in re mechanica abutuntur.

   Acta Eruditorum, March 1686.

   Reprinted in his Mathematische Schriften. Halle, 1860; Hildesheim, 1962. Bd. 6.

   English translation "A brief demonstration of a notable error of Descartes and others concerning a natural law, according to which God is said always to conserve the same quantity of motion; a Law which they also misuse in mechanics" in L.E. Loemker. Leibniz' Philosophical Papers and Letters. Chicago: University of Chicago Press, 1956.

2. Essay de Dynamique sur les loix du mouvement, ou il est montré qu'il ne se conserve pas la même quantité de mouvement, mais la même force absolue, ou bien la même quantité de l'action motrice.

   Written in 1691; first pub. in 1860 in Mathematische Schriften von G.W. Leibniz. Berlin: Gerhardt, 1860. Vol. VI, pp. 215-231. Reprinted, Hildesheim, 1962.

John Edward LENNARD-JONES (1894-1954)

1. On the Determination of Molecular Fields. I. From the Variation of the Viscosity of a Gas with Temperature. II. From the Equation of State of a Gas. III. From Crystal Measurements and Kinetic Theory Data.

   Proc. Roy. Soc. London, Vol. A 106, pp. 441-462, 463-477, 709-718 (1924).

2. Cohesion.

   Proceedings of the Physical Society of London, Vol. 43, pp. 461-482 (1931).

Milton Stanley LIVINGSTON (1905-)

See E.O. LAWRENCE.

Mikhail Vasil'evich LOMONOSOV (1711-1765)

1. Meditationes de caloris et frigoris causa.

   Presented at a session of the Academy of Sciences, St. Petersburg, Dec. 7, 1744.

   Novi Commentarii Academiae scientiarum imperialis Petropolitanae, Vol. 1, pp. 206-229 (1750).

   Russian translation 1828.

   Excerpts in German trans. from the Russian by B.N. Menshutkin, in Ostwald's Klassiker, #178 (1910).

   Full Latin text with Russian translation given in Lomonosov Polnoe Sobrannie Sochinenii. Moscow and Leningrad: Academy of Sciences, 1951-1959.

   English translation by Henry M. Leicester in Mikhail Vasil'evich Lomonosov on the Corpuscular Theory. Cambridge: Harvard University Press, 1970.

2. Tentamen theoriae de vi aëris elastica.

   Presented to the Academy of Sciences, St. Petersburg, Sept. 2, 1748.

   Novi Commentarii Academiae scientiarum imperialis Petropolitanae, Vol. 1, pp. 230-244 (1750).

   Trans. into Russian by B.M. Menshutkin in 1936.

   Latin text and Russian translation reprinted in Lomonosov Polnoe Sobrannie Sochinenii.

   English translation in Henry M. Leicester. Mikhail Vasil'evich Lomonosov on the Corpuscular Theory.

Fritz LONDON (1900-1954)

1. Superfluids. Volume One: Macroscopic Theory of Superconductivity. New York: Wiley, 1950; 2nd ed. 1961; Dover reprint. Volume Two: Macroscopic Theory of Superfluid Helium. New York: Wiley, 1954.

Hendrik Antoon LORENTZ (1853-1928)

1. Versuch einer Theorie der elektrischen und optischen Erscheinungen in bewegten Körpern.

   Leiden, 1895. 2nd ed. (reprint), Leipzig: Teubner, 1906.

   An extract of 5 pages, proposing the contraction hypothesis to explain the Michelson-Morley experiment, is translated by W. Perrett and G.B. Jeffery in The Principle of Relativity by A. Einstein et al. London: Methuen, 1923. Reprinted by Dover Publications.

2. Electromagnetische verschijnselen in een stelsel, dat zich met willekeurige snelheid kleiner dan die van het licht beweegt.

   Verslagen Wis. Nat. Afd. K. Akad. Wet. Amsterdam, Ser. 4, Vol. 12, pp. 986-1009 (1904).

   English trans. "Electromagnetic Phenomena in a System Moving with Any Velocity Less than That of Light." Proceedings of the Academy of Sciences, Amsterdam, Vol. 6, pp. 809-831 (1904).

   Reprinted in The Principle of Relativity by H.A. Lorentz et al. London: Methuen & Co., Ltd., 1923; reprinted by Dover; and in C.W. Kilmister. Special Theory of Relativity. Oxford & New York: Pergamon Press, 1970.

3. The Theory of Electrons and Its Applications to the Phenomena of Light and Radiant Heat. Lectures at Columbia University. New York: Stechert/Leipzig: Teubner, 1909; 2nd ed., 1916; reprinted by Dover Publications, New York, 1952.

Josef LOSCHMIDT (1821-1895)

1. Zur Grösse der Luftmolecüle.

   Sitzungsber. Akad. Wiss. Wien, Abt. 2, Vol. 52, pp. 395-413 (1865).

Titus LUCRETIUS Carus (95? - 55? B.C.)

1. De Rerum Natura libri sex.

English translations by H.A.J. Munro (Cambridge, England, 1864); by W.H.D. Rouse (London and New York, 1924); and others. On the Nature of Things.

English translation by Rolfe Humphries. The Way Things Are. Bloomington, Ind.: Indiana University Press, 1969.

Ernst MACH (1838-1916)

1. Die Geschichte und die Wurzel des Satzes von der Erhaltung der Arbeit. Prague: Calve, 1872.

English translation by P.E.B. Jourdain. History and Root of the Principle of Conservation of Energy. Chicago: Open Court Pub. Co., 1911.

See also "On the Principle of Conservation of Energy" the Monist, Vol. 5, pp. 22-54 (1894). "Being in Part a Re-elaboration of the Treatise Ueber die Erhaltung der Arbeit," reprinted in Popular Scientific Lectures. Chicago: Open Court Pub. Co., 1895.

German edition: Populär wissenschaftliche Vorlesungen. Leipzig: J.A. Barth, 1896.

2. Die Mechanik in ihrer Entwicklung historisch-kritisch dargestellt. Leipzig: Brockhaus, 1883. 2nd through 9th editions, 1888, 1897, 1901, 1904, 1908, 1912, 1921, 1933.

English trans. by Thomas J. McCormack. The Science of Mechanics; 6th ed., with revisions through the 9th German ed. Chicago and LaSalle, Ill.: The Open Court Publishing Co., 1942, 1960.

French trans. (from 4th ed.), Paris: Hermann, 1904.

Ettore MAJORANA

1. Über die Kerntheorie.

Zeits. f. Physik, Vol. 82, pp. 137-145 (1933).

English translation "On Nuclear Theory" in D.M. Brink. Nuclear Forces. Oxford: Pergamon Press, 1965.

Etienne-Louis MALUS (1775-1812)

1. Sur une propriété de la lumiere réfléchie par les corps diaphanes.

Memoires Soc. Arcueil, Vol. 2, pp. 143-158 (1809).

German trans. in Ann.d.Physik, Vol. 31, pp. 286-296 (1809)

English trans. in Nicholson's J. Nat. Phil., Vol. 30, pp. 95-102 (1812)

Edme MARIOTTE (1620?-1684)

1. Essay de la nature de l'air. Paris, 1679.
   Reprinted in Mariotte's Oeuvres. Leiden, 1717; and the Hague, 1740.

E. MARSDEN

See H. GEIGER.

Robert Eugene MARSHAK (1916-)

See E.C.G. SUDARSHAN.

James Clerk MAXWELL (1831-1879)

1. Illustrations of the Dynamical Theory of Gases.
   Phil. Mag., Vol. 19, pp. 19-32; Vol. 20, pp. 21-37 (1860).
   Reprinted in The Scientific Papers of James Clerk Maxwell. Cambridge University Press, 1890; Dover Publications, 1965.

2. On Physical Lines of Force.
   Phil. Mag., Ser. 4, Vol. 21, pp. 161-175, 281-291, 338-348 (1861); Vol. 23, pp. 12-24, 85-95 (1862).
   Reprinted in his Scientific Papers.
   German trans. ed. by L. Boltzmann, Ueber Physikalische Kraftlinien. Leipzig: W. Engelmann, 1898; Ostwalds Klassiker nr. 102.

3. On Faraday's Lines of Force.
   Trans. Cambridge Phil. Soc., Vol. 10, pp. 27-83 (1864).
   Reprinted in his Scientific Papers.
   German trans. ed. L. Boltzmann, Ueber Faraday's Kraftlinien. Leipzig: W. Engelmann, 1895; Ostwalds Klassiker nr. 69.

4. On the Dynamical Theory of Gases.
   Phil. Trans. Roy. Soc. London, Vol. 157, pp. 49-88 (1867); Phil. Mag., Ser. 4, Vol. 32, pp. 390-393 (1866); Vol. 35, pp. 129-145, 185-217 (1868).
   Reprinted in his Scientific Papers.
   Reprinted in S.G. Brush. Kinetic Theory. Vol. 2. Oxford: Pergamon Press, 1966.

5. A Treatise on Electricity and Magnetism. Oxford: Clarenden Press, 1873; 2nd ed. 1881; 3rd ed. 1891.
   Reprinted by Dover Publications, New York, 1954.
   German trans. by M.B. Weinstein. Lehrbuch der Electricität und des Magnetismus. Berlin: Springer, 1883.

Julius Robert MAYER (1814-1878)

1. Bemerkungen über die Kräfte der unbelebten Natur.

   Ann. Chem. Pharm., Vol. 42, pp. 233-240 (1842).

   Reprinted in Die Mechanik der Wärme in gesammelten Schriften. Stuttgart, 1867.

   English translations by G.C. Foster. "The Forces of Inorganic Nature," Philosophical Magazine, Ser. 4, Vol. 24, pp. 371-377 (1862).

   Foster's translation reprinted in S.G. Brush. Kinetic Theory. Vol. 1. Oxford: Pergamon Press, 1965.

2. Die organische Bewegung in ihrem Zusammenhang mit dem Stoffwechsel. Ein Beitrag zur Naturkunde. Heilbronn, 1845.

   Reprinted in his Mechanik der Wärme. Stuttgart, 1867.

   English translation to be published by R.B. Lindsay in a book on Mayer. Oxford: Pergamon Press.

3. Bemerkungen über das mechanische Aequivalent der Wärme. Heilbronn & Leipzig, 1851.

   Reprinted in Die Mechanik der Wärme.

   English trans. by G.C. Foster. "Remarks on the Mechanical Equivalent of Heat" in Phil. Mag., Series 4, Suppl. to Vol. 25, pp. 493-522 (1863).

B.V. MEDVEDEV

See N.N. BOGOLIUBOV.

Albert Abraham MICHELSON (1852-1931)

1. The Relative Motion of the Earth and the Luminiferous Ether. Amer. J. Sci., Series 3, Vol. 22, pp. 20- (1881).

   Reprinted in C.W. Kilmister. Special Theory of Relativity. New York and Oxford: Pergamon Press, 1970.

Albert Abraham MICHELSON and Edward W. MORLEY

2. On the Relative Motion of the Earth and the Luminferous Ether. Amer. J. Sci., Series 3, Vol. 34, pp. 333-345 (1887); Phil. Mag., Series 5, Vol. 24, pp. 449-463 (1887).

Hermann MINKOWSKI (1864-1909)

1. Die Grundgleichungen für die elektromagnetischen Vorgänge in bewegten Körpern.

   Nachr. Ges. Wiss. Göttingen, pp. 53-111 (1908).

2. Raum und Zeit.

   Physik. Z., Vol. 10, pp. 104-111 (1909).

   English translation by W. Perrett and G.B. Jeffery in "Space and Time," The Principle of Relativity by H.A. Lorentz, A. Einstein, H. Minkowski, and H. Weyl. London: Methuen, 1923; reprinted by Dover.

Edward Williams MORLEY (1838-1923)

See A.A. MICHELSON.

Henry Gwyn Jeffreys MOSELEY (1887-1915)

1. The High-Frequency Spectra of the Elements.

   Phil. Mag., Series 6, Vol. 26, pp. 1024-1034 (1913); and Vol. 27, pp. 703-713 (1914).

   Reprinted in Stephen Wright. Classical Scientific Papers-Physics. New York: American Elsevier Pub.Co., 1965.

Claude Louis Marie Henri NAVIER (1785-1836)

1. Mémoire sur les lois du mouvement des fluides.

   Mem. Acad. Sci. Paris (Ser. 2), Vol. 6, pp. 289-440 (1882, pub. 1827).

Walter Hermann NERNST (1864-1941)

1. Über die Berechnung chemischer Gleichgewichte aus thermischen Messungen.

   Nachr. Kgl. Ges. d. Wiss., Göttingen, Math.-Phys. Klasse, 1906, Heft 1, pp. 1-40.

Isaac NEWTON (1642-1727)

1. A New Theory about Light and Colours.

   Phil. Trans. Roy. Soc. London, No. 80, pp. 3075-3087 (Feb. 1672); Lowthorp's abridged Philosophical Transactions, 2nd ed., pp. 128-135.

   Reprinted by W. Fritsch. München, 1965.

2. Philosophiae Naturalis Principia Mathematica. London, 1687. 2nd ed. with a new preface by R. Cotes. Cambridge, 1713. 3rd ed., London, 1726.

   English translation by Andrew Motte. Sir Isaac Newton's Mathematical Principles of Natural Philosophy and His System of the World. London, 1729. Revised, and with historical explanatory appendix by Florian Cajori. Berkeley: University of California Press, 1934.

   Reprint of the 1st Latin edition of 1687 by Dawson's, 1960.

   French trans. by Marquise du Chastellet. Principes mathematiques de la philosophie naturelle. Paris, 1756, reprinted by A. Blanchard, Paris, 1966.

   See I.B. Cohen, "The French Translation of Isaac Newton's Philosophiae Naturalis Principia Mathematica (1756, 1759, 1966)," Arch. Int. Hist. Sci., Vol. 21, pp. 261-290 (1968).

   German trans. by J.P. Wolfers. Sir I. Newton's mathematische Principien der Naturlehre. Berlin, 1872.

3. Opticks; Or, a Treatise of the Reflections, Refractions, Inflections & Colours of Light. London, 1704.

   Latin translation by Samuel Clarke. Optice... (1706).

   4th ed. (London, 1730), reprinted by G. Bell (London, 1931) with a foreword by Albert Einstein, an introduction by Sir Edmund Whittaker; reprinted by Dover Pub., New York, 1952, with an additional Preface by I. Bernard Cohen and an analytical table of contents by Duane H.D. Roller.

   German trans. by W. Abendroth. Sir Isaac Newton's Optik ... Ostwald's Klassiker nr. 96-97. Leipzig, 1898.

Hans Christian OERSTED (1777-1851)

1. Experimenta Circa Effectum Conflictus Electrici in Acum Magneticam. Hafniae, 1820.

   English translation, "Experiments on the Effect of a Current of Electricity on the Magnetic Needle," Annals of Philosophy, Vol. 16, pp. 273-276 (1820).

   English trans. reprinted in R.A.R. Tricker. Early Electrodynamics. Oxford: Pergamon Press, 1965.

   German trans. in Ann.d. Physik, Vol. 66, pp. 295-304 (1820).

   German translation."Versuche über die Wirkung des elektrischen Conflicts auf die Magnetnadel," in Abhandlungen von Hans Christian Oersted und Thomas Johann Seebeck. Leipzig: Engelmann, 1895 (Ostwald's Klassiker, nr. 63).

French translation, "Experiences sur l'effect du conflict electrique sur l'aiguille aimantêe" in Annales de Chimie et de Physique, Vol. 14, pp. 417-426 (1820).

Facsimiles of the first Latin, French, Italian, German, English and Danish announcements reprinted in La Decouverte de l'Electromagnetisme faite en 1820 par J.-C. Oersted. Copenhagen, 1920.

Georg Simon OHM (1789-1854)

1. Die Galvanische Kette, mathematisch bearbeitet. Berlin: T.H. Riemann, 1827.

   English translation by William Francis. The Galvanic Circuit Investigated Mathematically. New York: D. Van Nostrand, 1891. Kraus Reprint Co., New York, 1969.

Lars ONSAGER (1903-)

1. Reciprocal Relations in Irreversible Processes.

   Phys. Rev., Series 2, Vol. 37, pp. 405-426; Vol. 38, pp. 2265-2279 (1931).

2. Crystal Statistics. I. A Two-Dimensional Model with an Order-Disorder Transition.

   Phys. Rev., Series 2, Vol. 65, pp. 117-149 (1944).

Leonard Salomon ORNSTEIN (1880-) and Fritz ZERNIKE

1. De toevallige dichtheidsafwykingen en de opalescentie by het kritisch punt van eene enkelvoudige stof.

   Verslagen Wis. Nat., Afd. K. Akad. Wet. Amsterdam, Vol. 23, pp. 582-595 (1914).

   English trans. "Accidental Deviations of Density and Opalescence at the Critical Point of a Single Substance" in Proc. Acad. Sci. Amsterdam, Vol. 17, pp. 793-806 (1914).

   English trans. reprinted in H.L. Frisch & J.L. Lebowitz. The Equilibrium Theory of Classical Fluids. New York: Benjamin, 1964.

L.S. ORNSTEIN

See also G.E. UHLENBECK.

Blaise PASCAL (1623-1662) and Florin PERIER

1. Récit de la grande expérience de l'equilibre des liqueurs, projetée par le sieur B.P. pour l'accomplissement du traité qu'il a promis dans son abrégé touchant la vide et faite par le sieur F.P. en une des plus hautes montagnes d'Auvergne. Savreuz, Paris, 1648.

   Reprinted in Pascal's Oeuvres. Ed., L. Brunschvieg & Boutroux. Paris, 1908-1921, Vol. 2; also in Oeuvres Completes. Ed. J. Chevalier. Paris, 1954.

   English translation in The Physical Treatises of Pascal, The Equilibrium of Liquids and the Weight of the Mass of the Air. Trans. by I.H.B. and A.G.H. Spiers, with introduction and notes by F. Barry. New York: Columbia University Press, 1937. This includes related writings by Simon Stevin, Galileo Galilei and Torricelli. Spiers trans. reprinted in Peter Wolff. Breakthroughs in Physics. New York: The New American Library, 1965 (Signet Science Library).

Blaise PASCAL

2. Traités de l'équilibre des liqueurs et de la pesanteur de la masse de l'air. Paris: G. Despretz, 1663 (written 1653?).

   English trans. by I.H.B. and A.G.H. Spiers in The Physical Treatises of Pascal: The Equilibrium of Liquids and the Weight of the Mass of the Air. New York: Columbia University Press, 1937.

Wolfgang PAULI (1900-1958)

1. Über den Zusammenhang des Abschlusses der Elektronengruppen im Atom mit der Komplexstruktur der Spektren.

   Z.f. Physik, Vol. 31, pp. 765-783 (1925).

   English translation "On the Connexion Between the Completion of Electron Groups in an Atom with the Complex Structure of Spectra" in D. ter Haar. The Old Quantum Theory. Oxford: Pergamon Press, 1967 (some footnotes omitted).

2. The Connection Between Spin and Statistics.

   Phys. Rev., Series 2, Vol. 58, pp. 716-722 (1940).

   Reprinted in J. Schwinger. Quantum Electrodynamics. New York: Dover Publications, 1958.

Jean Charles Athanase PELTIER (1785-1845)

1. Nouvelles expériences sur la caloricité des courants électriques. Ann. Chim. Phys., Vol. 56, pp. 371-386 (1834).

Jean Baptiste PERRIN (1870-1942)

1. L'agitation moleculaire et le mouvement brownien.

   Compt. rend. Acad. Sci. Paris, Vol. 146, pp. 967-970 (1908).

   Reprinted in Oeuvres Scientifiques de Jean Perrin. Paris: Centre National de la Recherche Scientifique, 1950.

2. L'origine du mouvement brownien.

   Compt. rend. Acad. Sci. Paris, Vol. 147, pp. 530-532 (1908).

3. Grandeur des molecules et charge de l'elèctron.

   Compt. rend. Acad. Sci. Paris, Vol. 147, pp. 594-596 (1908).

4. Mouvement brownien et réalite moléculaire.

   Annales de Chimie et de Physique, Series 8, Vol. 18, pp. 1-114 (1909).

   Reprinted in his Oeuvres.

   German trans. "Die Brownsche Bewegung und die wahre Existenz der Moleküle" in Kolloid-chem-Beih. Vol. 1. pp. 221-300. Dresden, 1910.

   English trans. Brownian Movement and Molecular Reality.

Max Karl Ernst Ludwig PLANCK (1858-1947)

1. Ueber eine Verbesserung der Wien'schen Spectralgleichung.

   Verh. D. Phys. Ges., Vol. 2, pp. 202-204 (1900).

   Reprinted in Planck's Physikalische Abhandlungen und Vorträge. Braunschweig: Vieweg, 1958.

   Reprinted in Die Ableitung der Strahlungsgesetze. Leipzig, 1923. Ostwald's Klassiker, nr. 206.

   English translation "On An Improvement of Wien's Equation for the Spectrum" in D. ter. Haar. The Old Quantum Theory. Oxford: Pergamon Press, 1967.

2. Zur Theorie des Gesetzes der Energieverteilung im Normalspectrum.

   Verh. D. Phys. Ges., Vol. 2, pp. 237-245 (1900).

   Reprinted in Planck's Physikalische Abhandlungen und Vorträge.

   Reprinted in Die Ableitung der Strahlungsgesetze. Leipzig, 1923 (Ostwald's Klassiker, nr. 206).

English trans. "On the Theory of the Energy Distribution Law of the Normal Spectrum" in D. ter Haar. The Old Quantum Theory.

3. Vorlesungen über die Theorie der Wärmestrahlung.

   Leipzig: J.A. Barth, 1906; 2nd to 5th eds. 1913, 1919, 1921, 1923.

   English trans. by Morton Masius from the 2nd German ed. The Theory of Heat Radiation. Philadelphia, 1914; reprinted by Dover Publications, Inc., New York, 1959.

Henri POINCARE (1854-1912)

1. Sur la dynamique de l'electron.

   Rendiconti del Circolo mat di Palermo, Vol. 21, pp. 129-179 (1906).

   Reprinted in Poincare's Oeuvres. Vol. 9, pp. 494-550. Paris: Gauthier-Villars, 1951-1954.

   Partial English translation in "The Dynamics of the Electron," in C.W. Kilmister. Special Theory of Relativity. Oxford & New York: Pergamon Press, 1970.

Siméon Denis POISSON (1781-1840)

1. Mémoire sur la Théorie du Son.

   J. École Polytech., 14eme Cah., Vol. 7, pp. 319-392 (1808).

2. Mémoire su le mouvement des fluides élastiques dan les tuyaux cylindriques, et sur la théorie des instruments à vent.

   Mém. Acad. Roy. Sci. Inst. France, année 1817, Vol. 2, pp. 305-402 (1818).

3. Mémoire sur l'intégration de quelques équations linéaires aux différences partielles, et particulièrement de l'équations générale du mouvement des fluides élastiques.

   Mém. Acad. Roy. Sci. Inst. France, année 1817, Vol. 3, pp. 121-176 (1820).

4. Mémoire sur l'equilibre et le mouvement des corps elastiques.

   Mém. Acad. Roy. Sci. Inst. France, Series 2, Vol. 8, pp. 357-570 (1829).

M.K. POLIVANOV

See N.N. BOGOLIUBOV.

Charles Edwin PORTER (1926-1964)

See H. FESHBACH.

Joseph PRIESTLEY (1733-1804)

1. The History and Present State of Electricity, with Original Experiments. London: J. Dodsley, 1767; 2nd ed., 1769; 3rd ed., 1775; 4th ed., 1775.

   Reprinted with appendix and introduction by R. Schofield. New York: Johnson Reprint Corp., 1966.

   German trans. by J.G. Krünitz. Geschichte und gegenwärtiger Zustand der Electricität. Berlin, 1772.

   French trans. by M.J. Brisson. Histoire de l'électricité. Paris, 1771.

John William Strutt, Third Baron RAYLEIGH (1842-1917)

1. On the Light from the Sky, Its Polarization and Colour.

   Phil. Mag., Series 4, Vol. 41, pp. 107-120, 274-279 (1871).

   Reprinted in his Scientific Papers. Cambridge University Press, 1899-1920; reprinted by Dover Publications, New York, 1964.

2. The Theory of Sound. 2 vols. London: Macmillan, 1877-1878; 2nd ed., 1894; reprinted by Dover Publications, Inc., New York, 1945.

3. Remarks upon the Law of Complete Radiation.

   Phil. Mag., Ser. 49, pp. 539-540 (1900).

   Reprinted in his Scientific Papers.

Andrew REID

See G.P. THOMSON.

Frederick REINES (1918-   ) and C.L. COWAN, Jr.

1. The Detection of the Free Neutrino.

   Phys. Rev., Series 2, Vol. 92, p. 830 (1953).

Robert C. RETHERFORD (1912-)

See W.E. LAMB, Jr.

Owen Williams RICHARDSON (1879-1959) and Karl T. COMPTON

1. The Photoelectric Effect.

   Phil. Mag., Series 6, Vol. 24, pp. 575-594 (1912).

   Reprinted in Stephen Wright. Classical Scientific Papers——Physics. New York: American Elsevier Pub. Co., 1965.

Olaus RØMER (1644-1710)

1. Demonstration touchant le mouvement de la lumière trouvé par M. Rømer.

   Journal des Scavans, Dec. 7, 1676, pp. 233-236.

   English version (not an exact translation), "A Demonstration Concerning the Motion of Light" in Phil. Trans. Roy. Soc. London, No. 136, pp. 893-894 (June 1677).

   Both reprinted by I.B. Cohen. "Roemer and the First Determination of the Velocity of Light (1676)," Isis, Vol. 31, pp. 327-379 (1940).

Wilhelm Conrad RÖNTGEN (1845-1923)

1. Ueber eine neue Art von Strahlen.

   Sitzungsberichte der Würzburger physikalische-medicinischen Gesellschaft, pp. 132-141 (1895); pp. 11-16, 17-19 (1896).

   (Partial?) English trans. "On a New Type of Rays" in X-Rays and the Electric Conductivity of Gases. Alembic Club Reprints No. 22. Edinburgh: E. & S. Livingstone Ltd., 1958.

T. ROYDS

See E. RUTHERFORD.

Benjamin Thompson, Count RUMFORD (1753-1814)

1. An Inquiry Concerning the Source of the Heat Which Is Excited by Friction.

   Phil. Trans. Roy. Soc. London, Vol. 88, pp. 80-102 (1798).

Reprinted in his Essays, Political, Economical, and Philosophical. New ed. (London, 1800), Vol. 2; in his Complete Works (Boston, 1870-75), Vol. 2; and in Collected Works of Count Rumford. Ed. S.C. Brown. Harvard University Press, 1968.

Ernest RUTHERFORD (1871-1937)

1. A Radio-Active Substance Emitted from Thorium Compounds.

   Phil. Mag., Series 5, Vol. 49, pp. 1-14 (1900).

   Reprinted in Alfred Romer. The Discovery of Radioactivity and Transmutation. New York: Dover Publications, 1964.

2. Radioactivity Produced in Substances by the Action of Thorium Compounds.

   Phil. Mag., Series 5, Vol. 49, pp. 161-192 (1900).

   Reprinted in Alfred Romer. The Discovery of Radioactivity and Transmutation.

Ernest RUTHERFORD and Frederick SODDY

3. The Cause and Nature of Radioactivity.

   Phil. Mag., Series 6, Vol. 4, pp. 370-396 (1902).

   Reprinted in Stephen Wright. Classical Scientific Papers—Physics. New York: American Elsevier Pub. Co., 1965.

4. The Radioactivity of Thorium Compounds. I. An Investigation of the Radioactive Emanation. II. The Cause and Nature of Radioactivity.

   J. Chem. Soc., Trans. Vol. 81, pp. 321-350, 837-860 (1902).

   Reprinted in Alfred Romer. The Discovery of Radioactivity and Transmutation.

Ernest RUTHERFORD

5. The Magnetic and Electric Deviation of the Easily Absorbed Rays from Radium.

   Phil. Mag., Series 6, Vol. 5, pp. 177-187 (1903).

   Reprinted in Wright. Classical Scientific Papers—Physics.

Ernest RUTHERFORD and Frederick SODDY

6. Radioactive Change.

Phil. Mag., Series 6, Vol. 5, pp. 576-591 (1903).

Reprinted in Alfred Romer. The Discovery of Radioactivity and Transmutation. New York: Dover Pubs., 1964.

Ernest RUTHERFORD

7. The Succession of Changes in Radioactive Bodies. Bakerian Lecture, May 19, 1904.

   Phil. Trans. Roy. Soc. of London, Vol. A 204, pp. 169-219 (1905).

   Reprinted in Alfred Romer. The Discovery of Radioactivity and Transmutation.

8. The Mass and Velocity of the Alpha Particles Expelled from Radium and Actinium.

   Phil. Mag., Series 6, Vol. 12, pp. 348-371 (1906).

   Reprinted in Wright. Classical Scientific Papers—-Physics.

Ernest RUTHERFORD and T. ROYDS

9. The Nature of the Alpha Particle from Radioactive Substances.

   Phil. Mag., Series 6, Vol. 17, pp. 281-286 (1909).

   Reprinted in Wright. Classical Scientific Papers—-Physics.

Ernest RUTHERFORD

10. The Scattering of α and β Particles by Matter and the Structure of the Atom.

    Phil. Mag., Series 6, Vol. 21, pp. 669-688 (1911).

    Reprinted in R.T. Beyer. Foundations of Nuclear Physics. New York: Dover Publications, Inc., 1949.

    Reprinted in D. ter Haar. The Old Quantum Theory. Oxford: Pergamon Press, 1967.

Ernest RUTHERFORD and J.M. NUTTALL

11. Scattering of Alpha Particles by Gases.

    Phil. Mag., Series 6, Vol. 26, pp. 702-712 (1913).

    Reprinted by Wright. Classical Scientific Papers---Physics.

Ernest RUTHERFORD

12. The Structure of the Atom.
 Phil. Mag., Series 6, Vol. 27, pp. 488-498 (1914).
 Reprinted in Wright. Classical Scientific Papers—Physics.

13. Collision of α Particles with Light Atoms. Part IV: An Anomalous Effect in Nitrogen.
 Phil. Mag., Series 6, Vol. 37, pp. 581-587 (1919).
 Reprinted in R.T. Beyer. Foundations of Nuclear Physics. New York: Dover Publications, Inc., 1949.

14. Nuclear Constitution of Atoms.
 Proc. Roy. Soc. London, Vol. A 97, pp. 374-400 (1920).
 Reprinted in Wright. Classical Scientific Papers—Physics.

See also J.J. THOMSON.

Otto SACKUR (1880-1914)

1. Die Anwendung der kinetischen Theorie der Gase auf chemische Probleme.
 Ann. d. Physik, Series 4, Vol. 36, pp. 958-980 (1911).

2. Die Bedeutung des elementaren Wirkungsquantums für die Gastheorie und die Berechung der chemischen Konstanten.
 Festschrift W. Nernst, zu seinem Fünfundzwanzigjähren Doktorjubilaum, gewidmet von seinen Schülern. Halle a.d.S.: Verlag von Wilhelm Knapp, 1912 (pp. 405-423).

Felix SAVART (1791-1841)

See J.B. BIOT.

Joseph SAUVEUR (1653-1710)

1. Système génerale des intervalles des sons.
 Mem. Acad. Roy. Sci. Paris, 1701, p. 297-.

Paul SCHERRER (1890-)

See P.J.W. DEBYE.

John Robert SCHRIEFFER (1931-)

See J. BARDEEN.

Erwin SCHRÖDINGER (1887-1961)

1. Quantisierung als Eigenwertproblem.
 Ann. d. Physik, Ser. 4, Vol. 79, pp. 361-376, 489-527, Vol. 80, pp. 437-490, Vol. 81, pp. 109-139 (1926).
 Reprinted in his Abhandlungen zur Wellenmechanik. Leipzig: J.A. Barth, 1926. 2nd ed. 1928.

Eng. trans. (from the 2nd German ed. of Abhandlungen) by J.F. Shearer and W.M. Deans. Collected Papers on Wave Mechanics. London: Glasgow, 1928.

Partial English translation in G. Ludwig. Wave Mechanics. Oxford: Pergamon Press, 1968.

French translation by A. Proca. Mémoires sur la Mécanique Ondulatoire. Paris: F. Alcan, 1933.

2. Über das Verhältnis der Heisenberg-Born-Jordanschen Quantenmechanik zu der meinen.

   Annalen der Physik, Series 4, Vol. 79, pp. 734-756 (1926).

   Reprinted in his Abhandlungen zur Wellenmechanik.

   English translation in Collected Papers on Wave Mechanics.

   English translation in G. Ludwig, Wave Mechanics.

   French translation in Mémoires sur la Mécanique Ondulatoire.

Julian SCHWINGER (1918-)

1. Quantum Electrodynamics.

   Phys. Rev., Series 2, Vol. 74, pp. 1439-1461 (1948); Vol. 75, pp. 651-679 (1949); Vol. 76, pp. 790-817 (1949).

   Third paper only is reprinted in J. Schwinger. Quantum Electrodynamics. New York: Dover Publications, 1958.

Thomas Johann SEEBECK (1770-1831)

1. Ueber den Magnetismus der galvanischen Kette.

   Abhandlungen der Königlichen Akademie der Wissenschaften (Berlin), pp. 289-346 (1820-1821).

   Reprinted in Zur Entdeckung des Electromagnetismus. Ed. A.J. v. Oettingen, Ostwald's Klassiker #70. Leipzig, 1895.

2. Magnetische Polarisation der Metalle und Erz durch Temperatur-Differenz.

   Abhandlungen der Königlichen Akademie der Wissenschaften (Berlin), pp. 265-373 (1822-1823).

   Reprinted in Ostwald's Klassiker #70. Leipzig, 1895.

Robert SERBER (1909-)

See S. FERNBACH.

Ja. G. SINAI

1. Ergodicity of Boltzmann's Gas Model.

   Statistical Mechanics: Foundations and Applications. Proceedings of the I.U.P.A.P. Meeting at Copenhagen, 1966). Ed. by Thor A. Bak. New York: Benjamin, 1967 (pp. 559-573).

   This is a summary of the author's work previously published in several papers in Russian, some of which have been translated into English, e.g., "On the Foundations of the Ergodic Hypothesis for a Dynamical System of Statistical Mechanics," Soviet Mathematics, Vol. 4, pp. 1818-1822 (1963).

John Clarke SLATER (1900-)

See N. BOHR.

Marian von SMOLUCHOWSKI (1872-1917)

1. Zarys kinetycznej teoriji ruchów Browna i roztworów metnych.

   Rozprawy i Sprawozdania z Posiedzen Wydzialu Matematyczno-Przyrodniczego Akademii Umiejetnošci (Krajow), Vol. A 46, pp. 257-282 (1906).

   German trans. in Annalen der Physik, Ser. 4, Vol. 21, pp. 756-780 (1906). "Zur kinetischen Theorie der Brownschen Molekularbewegung und der Suspensionen". Reprinted in Abhandlungen über die Brownsche Bewegung und verwandte Erscheinungen. Leipzig, 1923. Ostwald's Klassiker nr. 207.

   Partial English trans. by S.G. Brush, unpublished.

Frederick SODDY (1877-1956)

See E. RUTHERFORD.

Otto STERN (1888-)

See W. GERLACH.

Simon STEVIN (1548-1620)

1. The Principal Works of Simon Stevin. Ed. by the Stevin Committee set up by the Section of Science of the Royal Netherlands Academy. Original text and illustrations reproduced in facsimile; English translation with extensive critical notes on opposite pages. Amsterdam, 1955-1966. Vols. I-V.

George Gabriel STOKES (1819-1903)

1. On the Theories of the Internal Friction of Fluids in Motion, and of the Equilibrium and Motion of Elastic Solids.

   Trans. Cambridge Phil. Soc., Vol. 8, pp. 287-319 (1845).

   Reprinted in his Mathematical and Physical Papers. Vol. I. Cambridge University Press, 1880.

   Reprinted by Johnson Reprint Corp., New York, 1966.

2. On the Effect of the Internal Friction of Fluids on the Motion of Pendulums.

   Trans. Cambridge Phil. Soc., Vol. 9, pp. [8]-[106] (1850).

   Reprinted in his Mathematical and Physical Papers. Vol.III.

Ennackel Chandy George SUDARSHAN (1931-) and R.F. MARSHAK

1. Chirality Invariance and the Universal Fermi Interaction.

   Phys. Rev., Ser. 2, Vol. 109, pp. 1860-1862 (1958).

Theodore Brewster TAYLOR (1925-)

See S. FERNBACH.

Edward TELLER (1908-)

See G. GAMOW; H.A. JAHN.

George Paget THOMSON (1892-)

1. The Diffusion of Cathode Rays by Thin Films of Platinum.

   Nature, Vol. 120, p. 802 (1927).

George Paget THOMSON and Andrew REID

2. Diffraction of Cathode Rays by a Thin Film.

   Nature, Vol. 119, p. 890 (1927).

George Paget THOMSON

3. Experiments on the Diffraction of Cathode Rays.

   Proc. Roy. Soc. London, Vol. A 117, pp. 600-609 (1928)

Joseph John THOMSON (1856-1940) and Ernest RUTHERFORD

1. On the Passage of Electricity through Gases Exposed to Röntgen Rays.

   Phil. Mag., Vol. 42, pp. 392-407 (1896).

   Reprinted in X-Rays and the Electric Conductivity of Gases. Alembic Club Reprints No. 22. Edinburgh: E. & S. Livingstone Ltd., 1958.

Joseph John THOMSON

2. Cathode Rays.

   Phil. Mag., Series 5, Vol. 44, pp. 293-316 (1897).

   Reprinted in Stephen Wright. Classical Scientific Papers ---Physics. New York: American Elsevier Pub. Co., 1965.

3. On the Charge of Electricity Carried by the Ions Produced by Röntgen Rays.

   Phil. Mag., Series 5, Vol. 46, pp. 528-545 (1898).

   Reprinted by Wright. Classical Scientific Papers--Physics.

4. On the Masses of the Ions in Gases at Low Pressures.

   Phil. Mag., Series 5, Vol. 48, pp. 547-567 (1899).

   Reprinted by Wright. Classical Scientific Papers---Physics.

Sin-itiro TOMONAGA (1906-)

1. (Original paper in Japanese.) Bull. I.P.C.R. (Riken-iho), Vol. 22, pp. 545- (1943).

   English trans. "On a Relativistically Invariant Formulation of the Quantum Theory of Wave Fields" in Prog. Theoret. Phys., Vol. 1, No. 2, pp. 1-13 (1946).

   English version reprinted in J. Schwinger. Quantum Electrodynamics. New York: Dover Publications, 1958.

Evangelista TORRICELLI (1608-1647)

1. Esperienza dell'Argento vivo. (Correspondence with Ricci, June 11, 18, and 28, 1644.)

   In Torricelli's Opere. Vol. 3. Faenza, 1919, pp. 186-188, 193-195, 198-201.

   English translation by V. Cioffari in The Physical Treatises of Pascal. New York: Columbia University Press, 1937.

There is a more formal account published by the Accademia del Cimento: Saggi di naturali Esperienze fatte nell' Accademia del Cimento. Firenze, 1666. This is reprinted, together with the letters mentioned above, in Neudrucke von Schriften und Karten über Meteorologie und Erdmagnetismus. Ed., G. Hellmann. No. 7. Berlin, 1897.

G.E. UHLENBECK and S. GOUDSMIT

1. Ersetzung der Hypothese vom unmechanischen Zwang durch eine Forderung bezüglich des inneren Verhaltens jedes einzelnen Electrons.

   Naturwissenschaften, Vol. 13, pp. 953-954 (October 1925).

   An English version, "Spinning Electrons and the Structure of Spectra" was written December 1925 and was published in Nature, Vol. 117, p. 264 (1925), together with a short note by Niels Bohr. This is reprinted in H.A. Boorse & Lloyd Motz. The World of the Atom. New York: Basic Books, 1966.

   Also reprinted in W.R. Hindmarsh. Atomic Spectra. Oxford: Pergamon Press, 1967.

George Eugene UHLENBECK (1900-)

2. Over statistische Methoden in de Theorie der Quanta. Proefschrift, Leiden.

   's-Gravenhage: Martinus Nijhoff, 1927.

G.E. UHLENBECK and L.S. ORNSTEIN

3. On the Theory of the Brownian Motion.

   Phys. Rev., Series 2, Vol. 36, pp. 823-841 (1930).

   Reprinted in Nelson Wax. Selected Papers on Noise & Stochastic Processes. New York: Dover Publications, 1954.

Johannes Diderik VAN DER WAALS (1837-1923)

1. Over de continuiteit van den gas- en vloeistoftoestand. Proefschrift, Leiden, 1873. Published by A.W. Sijthoff, Leiden.

   German version (revised with new material), trans. by Friedrich Roth. Die Continuität des gasförmigen und flüssigen Zustandes. Leipzig: J.A. Barth, 1881.

   English trans. by R. Threlfall and J.F. Adair (from the German version). "The Continuity of the Liquid and Gaseous States of Matter" in Physical Memoirs, Vol. 1, Part 3. London: Taylor and Francis, for the Physical Society, 1890.

Alessandro VOLTA (1745-1827)

1. On the Electricity Excited by the Mere Contact of Conducting Substances of Different Kinds (in French).

   Phil. Trans. Roy. Soc. London, pp. 403-431 (1800).

   Reprinted in his _Opere_, Edizione Nazionale, Hoepli, Milano, 1918, Vol. I, p. 563.

   German trans. by A.J. von Oettingen. "Ueber die bei blosser berührung leitender Substanzen verschiedener Art erregte Elektricität" in _Untersuchungen über den Galvanismus 1796 bis 1800_. Leipzig: W. Engelman, 1900; Ostwalds _Klassiker_ nr. 118.

Theodore VON KÁRMÁN (1881-1963)

See M. BORN.

John VON NEUMANN (1903-1957)

1. _Mathematische Grundlagen der Quantenmechanik_.

   Berlin: Springer, 1932; reprinted by Dover Publications, 1943.

   English trans. by R.T. Beyer. _Mathematical Foundations of Quantum Mechanics_. Princeton: University Press, 1955.

2. Proof of the Quasi-Ergodic Hypothesis.

   Proc. Nat. Acad. Sci., USA, Vol. 18, pp. 70-82 (1932).

   Reprinted in R. Bellman (ed.). _A Collection of Modern Mathematical Classics; Analysis_. New York: Dover, 1961.

John WALLIS (1616-1703)

1. A summary account given by Dr. John Wallis of the General Laws of Motion.

   "The General Laws of Motion" translated from the Latin. Philosophical Transactions, No. 43, p. 864 (15 Nov. 1668). in I.B. Cohen. _Readings in the Physical Sciences_. Cambridge, Mass., 1966.

Ernest Thomas Sinton WALTON (1903-   )

See J.D. COCKCROFT.

John James WATERSTON (1811-1883)

1. On the Physics of Media That Are Composed of Free and Perfectly Elastic Molecules in a State of Motion.

   Proc. Roy. Soc. London, Vol. 5, p.604 (1846). Abstract only.

   Phil. Trans. Roy. Soc. London, Vol. 183A, pp. 5-79 (1893).

   Reprinted in The Collected Scientific Papers of John James Waterson. Ed. J.S.Haldane. Edinburgh: Oliver & Boyd, 1928.

Victor Frederick WEISSKOPF (1908-)

See H. FESHBACH.

Eugene Paul WIGNER (1902-)

1. Über die elastischen Eigenschwingungen symmetrischer System.

   Nachr. Ges. Wiss., Göttingen, pp. 133-146 (1930).

   English translation, "The Elastic Characteristic Vibrations of Symmetrical Systems" in Arthur P. Cracknell. Applied Group Theory. Oxford: Pergamon Press, 1968.

See also G. BREIT.

Charles Thomson Rees WILSON (1869-1959)

1. Condensation of Water Vapour in the Presence of Dust-Free Air and Other Gases.

   Phil. Trans. Roy. Soc. London,Vol. A 188, pp.265-307 (1897).

   Reprinted in Stephen Wright. Classical Scientific Papers—Physics. New York: American Elsevier Pub. Co., 1965.

2. On An Expansion Apparatus for Making Visible the Tracks of Ionising Particles in Gases and Some Results Obtained by Its Use.

   Proc. Roy. Soc. London, Vol. A 87, pp. 277-292 (1912).

   Reprinted in Wright. Classical Scientific Papers—Physics.

Marjorie WILSON and H.A. WILSON

1. On the Electric Effect of Rotating a Magnetic Insulator in a Magnetic Field.

   Proceedings of the Royal Society of London, Series A, Vol. 89, p. 99 (1913).

   Reprinted in C. W. Kilmister. Special Theory of Relativity. Oxford & New York: Pergamon Press, 1970.

Christopher WREN (1632-1723)

1. Lex naturae de collisione corperum.

   Phil. Trans. Roy. Soc. London, No. 43, p. 867 (1668).

   English trans., "The Law of Nature in the Collision of Bodies," in I.B. Cohen. Readings in the Physical Sciences. Cambridge, Mass., 1966.

Chien-Shiung WU (1913-), E. AMBLER, R.W. HAYWARD, D.D. HOPPES and R.P. HUDSON

1. Experimental Test of Parity Conservation in Beta-Decay.

   Phys. Rev., Ser. 2, Vol. 105, pp. 1413-1415 (1957).

   Reprinted in Charles Strachan. The Theory of Beta-Decay. Oxford: Pergamon Press, 1969.

Chen Ning YANG (1922-)

See T.D. LEE.

Thomas YOUNG (1773-1829)

1. On the Theory of Light and Colours.

   Phil. Trans. Roy. Soc., Vol. 92, pp. 12-48 (1802).

2. An Account of Some Cases of the Production of Colours, Not Hitherto Described.

   Phil. Trans. Roy. Soc., Vol. 92, pp. 387-397 (1802).

3. A Course of Lectures on Natural Philosophy and the Mechanical Arts. London, 1807. 2nd ed. (by P. Kelland), 1845.

Hideki YUKAWA (1907-)

1. On the Interaction of Elementary Particles I.

   Proceedings of the Physico-Mathematical Society of Japan (3) Vol. 17, pp. 48-57 (1935).

   Reprinted in R.T. Beyer. Foundations of Nuclear Physics. New York: Dover Publications, Inc., 1949.

   Reprinted in H.A. Boorse and Lloyd Motz. The World of the Atom. New York: Basic Books, 1966.

Fritz ZERNIKE

See L.S. ORNSTEIN.

4. WORKS NOT PRESENTLY AVAILABLE IN ENGLISH

The numbers refer to items listed in the previous section. An asterisk indicates that a translation needs to be completed, or an old translation needs to be reprinted, or a new translation may be in progress.

Aepinus	1	Guericke	1
d'Alembert	1, 2	Heisenberg	3*
Amontons	1, 2*	Huygens	2, 5*
Ampere	1,* 2	Ising	1
Boltzmann	2	Joliot-Curie	1
Born & von Kármán	1, 2, 3	Kepler	1
Born	4	Koenig	1
Born & Jordan	5*	Lagrange	1, 2
Born	6	Langevin	1
Born & Mayer	10	Lavoisier & Laplace	1
Born	11	Leibniz	2
de Broglie	1*	Lorentz	1*
Caratheodory	1,* 3	Loschmidt	1
Chladni	1	Mach	1*
Coulomb	1	Malus	1*
Descartes	2*	Mariotte	1
Duhem	1, 2	Mayer	2,* 3*
Einstein	10*	Minkowski	1
Euler	1, 2, 3, 4, 5*	Navier	1
Feddersen	1	Nernst	1
Fermi	1, 2*	Ornstein & Zernike	1*
Fizeau & Breguet	1	Peltier	1
Foucault	1	Perrin	1, 2, 3
Fourier	1, 2*	Poincare	1*
Fresnel	1, 2*	Poisson	1, 2, 3, 4
Gamow	1	Röntgen	1*
Goethe	1*	Sackur	1, 2
Grimaldi	1	Sauveur	1

Seebeck	1, 2	Uhlenbeck	2
Smoluchowski	1*	Volta	1

Similar lists of works not presently available in French, German, Italian, Japanese, Russian, and other languages should be made. We have not done this at this time because of our rather incomplete information about translations in these languages; but we plan to issue such lists in the future along with revisions in the list for works not presently available in English.

ADDENDA

Bernal, J. D. The Extension of Man, A History of Physics before the Quantum. London: Weidenfeld, 1971.

Burke, John G. Origins of the Science of Crystals. Berkeley: University of California Press, 1966.

Crosland, M. P., ed. The Science of Matter, A Historical Survey. Baltimore: Penguin Books, 1971. (P)

An anthology of 163 short extracts

Gillmor, C. S. Coulomb and the Evolution of Physics and Engineering in Eighteenth-Century France. Princeton, N. J.: Princeton University Press, 1971.

Hermann, A. et al. Lexikon Geschichte der Physik A-Z. Köln: Aulis Verlag Deubner & Co. KG, 1972.

"The contents include biographies of the important physicists, accounts of the development of the most significant methods and ideas, and the history of the most influential discoveries. There are about 500 separate subject headings... Each section is followed by an appendix giving references to relevant literature." (Publisher's announcement)

Metzger, Helene. La Genese de la Science des Cristaux. Paris: Albert Blanchard, 1969 (reprint of the 1918 edition).

Needham, Joseph. Science and Civilization in China. Vol. 4, Part I: Physics. Cambridge: Cambridge University Press, 1962.

Westfall, Richard S. Force in Newton's Physics, The Science of Dynamics in the Seventeenth Century. New York: American Elsevier 1971.

INDEX TO PARTS I AND II

INDEX TO PARTS I AND II

Wilson, C. T. R., II:74